Cold Facts About the Great Global Warming Scam

New Revelations that Decimate the Climate Change Narrative

Lynne Balzer

Faraday Science Institute

Dedication: Dr. Timothy Ball

Dr. Timothy Ball was a world-acclaimed climate scientist and lecturer. He was born in Chippenham, England on November 5, 1938. At the age of 19 he moved to Canada.

Tim Ball served in the Royal Canadian Air Force from 1960 to 1968.

Dr. Timothy Ball

In 1970 Dr. Ball received a B.S. degree from the University of Winnipeg. He received an M.A. degree from the University of Manitoba in 1971 and a PhD from Queen Mary University of London in 1983. Dr. Ball was a professor at the University of Winnipeg from 1971 to 1996. His area of expertise was historical climatology, including temperature reconstructions in Canada.

Dr. Ball authored several books, including *The Deliberate Corruption of Climate Science* in 2014 and *Human-Caused Global Warming: the Biggest Deception in History* in 2016. In addition, he contributed many articles to various publications and participated in public discussions and debates in climate science.

Dr. Ball received several awards, including the Clarence Atchison Award for Excellence in Community Service and the Clifford J. Robson Award for Teaching Excellence. He fought for the use of the scientific method to restore scientific integrity and to expose the pseudoscience being used to advance climate alarm.

Unfortunately, Dr. Ball was sued for over a million dollars by Penn State professor Michael Mann in 2011 for criticizing his fraudulent "hockey stick" graph that was published in 1998. This graph erroneously depicted world temperatures being essentially flat for the past 1,000 years — ignoring the Medieval Warm Period and the Little Ice Age — and then suddenly skyrocketing in the last decades of the twentieth century. Unfortunately that lawsuit dragged on for eight years, putting an immense strain on Dr. Ball's resources. Dr. Ball won the lawsuit in 2019 but only at great personal cost.

Dr. Ball passed away on September 24, 2022, and is sorely missed.

Foreword

The fact that you are reading this book indicates that you already suspect that something is very wrong with what we've been told about climate change. Although you may not know all of the reasons why these events have occurred, this book will provide you with a complete explanation of this controversial issue which has been plaguing us for nearly forty years.

Unfortunately, many people are **still** unaware that much of what they see and hear on TV, textbooks, magazines, newspapers and even Hollywood movies is just a series of carefully crafted stories behind an agenda those in power have in mind for us. These stories are being presented in a very convincing way: lies entwined with facts. The emergence of AI has made the job of untangling facts from lies all the more difficult. We also need to know how to distinguish facts from opinions. Emotions must be replaced by critical thinking. Ideas presented as facts must be checked against reality.

The internet provides a unique and priceless opportunity to find information to which previous generations never had access. This library of the world enables us to check claims from different perspectives. The bad news is that the internet also contains a lot of misinformation. Sifting through the facts and the misinformation takes time and skill. We may not always have this opportunity, since powerful forces are already closing some doors to the truth by blocking or removing material that they don't want us to see.

This book carefully lays out the facts about the claims of impending doom caused by human-caused (anthropogenic) global warming (AGW) or, as they say, "climate change" and the sustainability of the "solutions" being imposed on us without our permission.

Before allowing our lives to be taken over in an experiment to "prevent the world from ending", we need to learn the actual truth behind the climate change scare. Is the earth really headed for disaster if we fail to go along with the proposed solutions they've put in place to deal with this "problem"? Have our elected representatives – and the bureaucrats running the United Nations, the European Union and the Deep State – considered the consequences of such actions? Do they really believe we can save polar bears from extinction by changing to a different type of light bulb? Or that we can stop ocean levels from rising by building wind turbines? Or that we can make hurricanes less severe by driving electric cars? Are the thousands of news reports we've heard really accurate? Have we been told the truth?

This entire issue boils down to energy, which is the world's most important commodity. In the 19th century people were still using whale oil for lighting and burning wood to run steam engines. People relied on animals and slaves to produce farm crops. The discovery of petroleum elevated us to the modern age.

Without hydrocarbons, our civilization would grind to a halt. We would still be burning wood for heat and using animals for transportation. Can you imagine what would happen if we had to revert to this lifestyle? Yet that is what the global warming activists are asking us to do. The belief that we can just switch to photovoltaic panels and wind turbines is a pipe dream. They've been around for a while now and have demonstrated their unsustainability.

This updated book presents a comprehensive summary of the climate change fiasco in down-to-earth terms, beginning with the historical background — first global cooling and then global warming — starting in the 1960s, the incessant propaganda to which we've been subjected, and the looming specter of what will happen if the globalists are allowed to impose "the Great Reset" on the world.

There have been some encouraging signs that this insanity is sputtering to an end. One amazing development is that Bill Gates, always a climate change warrior, has now admitted that the world is not in a climate crisis. And now the Trump administration has taken the United States out of UN climate change organizations: The United Nations Framework Convention on Climate Change (UNFCCC) and the UN Intergovernmental Panel on Climate Change (IPCC). It is our earnest desire that other policymakers will follow his lead.

Hopefully the knowledge you acquire from reading this book will enable you to grasp the essential concepts you need to effectively communicate with those who are still in the dark about this crucial matter that concerns us all.

Contents

Chapter 1

Introduction

"We (UN-IPCC) redistribute de facto the world's wealth by climate policy. One has to free oneself from the illusion that international climate policy is environmental policy. This has almost nothing to do with environmental policy anymore." - Ottmar Endenhofer, IPCC Cochair of Working Group 3

Climate Countdown Clock

On the side of a building in Union Square in New York City is a giant digital clock that is four stories high and eighty feet wide. Ticking down the time until the climate change catastrophe occurs, this massive clock with orange-tinted digits serves to warn people that they'd better follow the dictates of the doomsayers if they want to save Planet Earth from being destroyed. When this clock strikes zero, they say, we will have arrived at an irreversible tipping point beyond which it will be impossible to avert this impending catastrophe.

Similar clocks exist in other places for the same purpose. After this one was built, other clocks synchronized with it began to appear in cities around the world.

The "experts" are telling us that limiting the earth's average temperature to 1.5° C. above the average global temperature in 1850, at the beginning of the Industrial Revolution, is critical for avoiding severe climate impacts. They maintain that exceeding this temperature increase will lead to extreme weather events, such as the disappearance of all the ice in the Arctic and uncontrolled sea level rise. One problem with this scenario is that there is no way to tell for sure what the earth's average temperature was in 1850. We can only guess.

Strangely, the countdown time on the Climate Clock is not definite. The time remaining until Doomsday can change. It will be affected by the amount of human emissions of carbon dioxide (CO_2) from burning hydrocarbons. If humanity is able to somehow reduce the amount of carbon dioxide we are emitting, the time until Doomsday will be extended. But if humans emit more carbon dioxide, it will be shortened. As of January, 2026, Doomsday on this countdown clock is **July 21, 2029**, a date that is shockingly close. Who decides what this date will be?

We've been repeatedly told that the earth is in an existential (threatening our existence) crisis that we humans have caused by burning hydrocarbons (known as "fossil fuels") such as coal, natural gas and petroleum. Water vapor, carbon dioxide and some other gases in the atmosphere, called "greenhouse gases", temporarily hold onto heat received from the sun, raising Earth's average temperature to a livable average, about 15° C (60° F). Were it not for this fortunate phenomenon, Earth's average temperature would be 33°C lower and life could not exist.

Everyone acknowledges that the earth is experiencing climate change. The climate has been changing since the beginning of Earth's history 4.5 billion years ago. Since the beginning of geologic history there have been approximately six great ice ages. We're now actually living between two parts of the Pleistocene Ice Age.

Based on data obtained from several proxies, an analysis of the temperatures of the last 450 million years of the earth's geologic history published in *Science Magazine* in September of 2024 by Judd et al reveals that we are living in the coldest period in the last 500 million years. Here is the graph published in this study. So the question is: "What is the ideal temperature for the planet?"

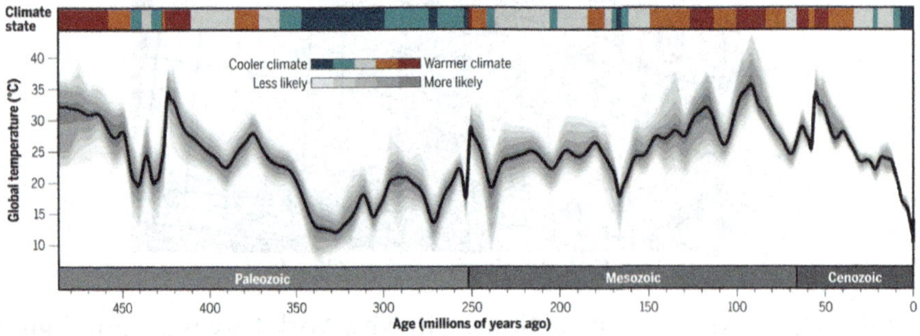

The two questions we need to ask about the climate are:

(1) "How much has the earth's temperature changed since the beginning of the Industrial Revolution?" (which they say is 1850) and

(2) "How much of this change is attributable to humans and how much of it to natural causes?"

Despite what you may have heard from academia and the mainstream media, the science has *not* been settled, despite the trillions of dollars that have now been spent on researching and dealing with this matter. Not only have taxpayers been saddled with an almost unbelievable expense, but — even worse — we are now witnessing the threat of losing basic freedoms of speech and movement being imposed upon us in the name of climate change.

Climate is average weather over a period of time, usually designated as a period of thirty or more years, and weather has a great influence on our daily activities. So everyone should understand basic facts about weather. Yet recently 50% of a group of Yale University students failed a test of basic climate knowledge. It's no wonder that most people feel inadequate to question what we're being told about changes in the climate and what we can do — if anything — about it.

The four main tenets of the human-caused (anthropogenic) global warming hypothesis are:

1) "Carbon dioxide [CO2] is the principal greenhouse gas."

This is false. Water vapor, making up about 2%–3% of the atmosphere, is the most powerful "greenhouse gas". At 20,000 parts per million (ppm), water vapor has 70-100 times as much effect on the warming of the atmosphere as carbon dioxide, which, at 420 ppm, makes up only about four hundredths of one percent (0.04%) of the atmosphere.

2) "Carbon dioxide controls global temperature, which has been steadily increasing."

To find out if this is true, we first need to examine previous weather reports to learn how much the average global temperature has increased in the last 175 years. We also need to know how much of this has been natural change and how much of it was caused by humans.

3) "Industrialized nations especially must atone, or pay the price, for this warming."

This makes it very clear that climate change is a political — not a scientific — issue.

4) "Man can control the climate. We can fix this problem by emitting less CO_2."

If it's true that mankind's addition of CO_2 to the atmosphere has caused climate change above and beyond natural causes, then theoretically we could stop it. But has it ever really been proved that carbon dioxide is the culprit?

Many people feel incompetent to investigate these questions on their own, relying instead on media they trust. The videos and articles presented by the global alarmists are scarce on evidence or citations. They tell us to rely on "experts". Please note that they rarely provide any proof to back up their pronouncements. They know that it's easier for the public to swallow these claims hook, line and sinker than to check them out using critical thinking, which asks the questions "Why?" and "How?"

A recent investigation revealed that most people feel uncomfortable with science and are thus reluctant to question this narrative. This is undoubtedly because too many of us were taught to view science as a long list of difficult equations and definitions rather than an exciting exploration of the world around us. The COVID "scamdemic", which revealed the failings of the official "science", awakened many of us to the reality of what has been happening in the last forty years.

You don't need a PhD in climate science to grasp the issue when the facts are presented in concrete, down-to-earth terms. Fortunately the Internet has made it easier than ever before to verify the accuracy of the anthropogenic global warming (AGW) hypothesis. Fifty years ago one would have had to comb through hundreds of books and periodicals in a huge library to get the same information that can now be obtained in just a few hours online.

But powerful forces have already closed some doors to the truth by censoring material that doesn't follow the official narrative — just as they've censored doctors warning us about COVID-19 mRNA "vaccines". So, to get to the truth, we have no time to lose in checking out the facts about this vital matter that concerns us all. Hopefully you will check the claims of impending doom caused by human-generated carbon dioxide and — almost as important — the probability that the proposed "renewables", which they say will alleviate the problem, will work.

Weather and climate are quite complex, and no one has all the answers. But this book will give you the facts you need for a thorough understanding of this critical issue and the dilemma in which we find ourselves.

Leo Tolstoy, author of *War and Peace,* observed that it's easier to explain a complicated thing to a simple person than a simple thing to an "educated" person who is brainwashed in dogma and believes s/he knows the answer.

We were told that 2024 was the hottest year in the last 125,000 years. Looking this up on line, we read, "The global average temperature for January-September 2024 was **1.54°** C above pre-industrial levels."

The earth's temperature did rise a bit in 2023 and 2024. In January of 2022 there was an underwater eruption of the Hunga-Tonga volcano in the South Pacific ocean. This eruption shot nearly three billion tons of water into the stratosphere and even the mesosphere, to an altitude of ~57 kilometers. Nothing quite like this has ever occurred in recorded history. "We have been fortunate to witness the largest climate event to occur on the planet since the advent of global satellite records, and possibly the largest event since the eruption of Mount Tambora in 1815."[1] The Mt. Tambora eruption caused the "year without a summer" in 1816. Hunga-Tonga's injection of water vapor, the main greenhouse gas, explains the delayed and short-lived temperature increase in 2023-2024.

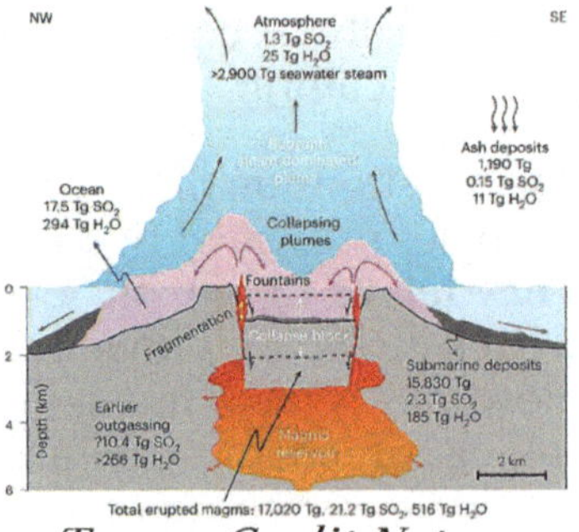

Hunga Tonga Credit:Nature-Geoscience

As a result of this eruption there was an extraordinary warming of the oceans beginning in late 2022 and extending into 2024. The temperature has since decreased.

However, to say that 2024 was the warmest year in 125,000 years is absolutely false, since it can be proven with proxy data that the Medieval Warm Period was about two degrees warmer than the present. It was also warmer during the 1930s Dust Bowl days. Temperatures were three to four degrees warmer during the Holocene Maximum around 8,000 years ago.

And how did they come up with a figure of 1.54°C? They couldn't have been using empirical science, which observes and measures the real world. In the 1800s most of the world was not recording daily temperatures. Looking at the map below, you can see that in 1850 the only places where temperatures were recorded daily were the eastern United States, western Europe and Australia. There were no weather stations taking daily temperature readings in most of South America, Africa, Asia or the polar regions. Temperature data were sparse or nonexistent.

The temperatures of the ocean, which makes up 70% of Earth's surface, were also guessed at.

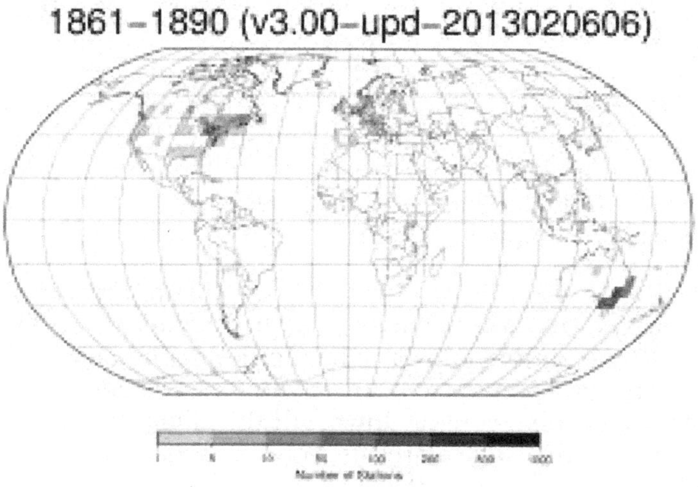

Weather Stations in the 1800s

Satellite temperature measurements, that would provide more accurate records, were not available until 1979. Ocean temperatures were not measured with any degree of accuracy until the establishment of the ARGO system in 2000.

Obtaining a mean (average) from such a lack of data would require a tremendous amount of guesswork. How in the world could the doomsayers now come up with a temperature increase to the **hundredth** of a degree (1.54 degrees)? Getting a figure this accurate would be impossible. The only way to obtain these readings was to just make them up.

Yet most people have accepted this claim without question. They've just assumed that the matter has been thoroughly explored using the scientific method. We've heard this claim repeated hundreds of times in so many ways in media reports and even in textbooks. How could all of these experts be wrong?

Critical thinking asks, "Is this true?" Most people won't even ask that question. Even journalists who have figured out that we are not in the midst of a "crisis" are afraid to openly question the assertion that Man has caused at least some unnatural heating.

Climate, average weather for an area over a thirty-year period, has previously been used to describe average weather for a region. An area near the ocean has a much different climate than a spot at the top of a mountain or a desert region. But climate scientists have attempted to average the temperature of the entire earth. So we will never get an accurate answer to the question, "How much has Earth's **average** temperature increased since the beginning of the Industrial Revolution around 1850?" This is the first lie.

The decade of the 1930s to 1940s was the warmest in recent experience. By the 1960s and 1970s, however, it was evident that temperatures were decreasing. Yet, at the same time the percentage of atmospheric CO_2 was rising. If CO_2, a product of combustion, causes global warming, then how could the earth cool down at the same time that CO_2 was increasing?

In the 1970s climatologists were telling the world that mankind was responsible for this **cooling.** On Earth Day of 1970 Dr. Kenneth Watt warned, "The world has been chilling sharply for about twenty years. If present trends continue, the world will be about four degrees colder for the global mean temperature in 1990, but eleven degrees colder in the year 2000. This is about twice what it would take to put us into an ice age."[2]

A *Washington Post* article published in 1971 warned, "In the next 50 years [2021] fine dust that humans discharge into the atmosphere by burning fossil fuel will screen out so much of the sun's rays that the earth's average temperature could fall by six degrees. Sustained emissions over five to ten years, could be sufficient to trigger an ice age."[3] (emphasis added)

Newsweek Magazine declared in April of 1975, "There are ominous signs that Earth's weather patterns have begun to change dramatically. . .The evidence in support of these predictions has now begun to accumulate so massively that meteorologists are hard-pressed to keep up with it. . .The central fact is that. . .the earth's climate seems to be cooling down. . .If the climate change is as profound as some of the pessimists fear, the results could be catastrophic."

In June of 1975, Nigel Calder, writing in *International Wildlife*, stated, "The continued rapid cooling of the earth since WWII is in accord with the increase in global air pollution associated with industrialization, mechanization, urbanization and exploding population." Articles about global cooling abounded in magazines, newspapers, TV reports and books. Some of the suggested solutions to this "crisis" included taking away our rights and freedoms. For example, Arnold Reitze, a professor at Case Western Reserve University, warned, "We will be forced to sacrifice democracy by the laws that will protect us from further pollution." He suggested outlawing internal combustion engines, placing rigid controls on all research and development in the production and marketing of new products and population controls, with a penalty imposed on couples producing more than the prescribed number of children.[4]

In April of 1975, *Newsweek Magazine* declared "There are ominous signs that Earth's weather patterns have begun to change dramatically. . .The evidence in support of these predictions has now begun to accumulate so massively that meteorologists are hard-pressed to keep up with it. . .The central fact is that. . .the earth's climate seems to be cooling down. . .If the climate change is as profound as some of the pessimists fear, the results could be catastrophic."

In his 1976 book, *The Cooling*, Lowell Ponte declared, "This cooling has already killed hundreds of thousands of people. If it continues and no strong action is taken, it will cause world famine, world chaos and world war, and this could all come about **before the year 2000.**" (emphasis added) Today's global warming alarmists are hoping that nobody remembers these dire predictions that were big news for nearly a decade. Trolls on social media are trying to convince us that this cooling scare never happened.

In 1971 NASA scientists estimated that if the percentage of carbon dioxide were doubled from about 0.03% to 0.06% of the atmosphere, the earth's average temperature would rise six to eight hundredths of a degree over a period of 1000 years, due to feedback from increased water vapor. In 1979 a team from the National Academy of Sciences concluded that a doubling of CO2 would cause no measurable temperature increase except for feedback from water vapor.

When the temperatures started to rise again in the early 1980s, the alarmists changed gears and — in lockstep — started issuing opposite predictions: this time that the earth was about to warm uncontrollably, with anthropogenic (human-caused) carbon dioxide still being the culprit.

Two Decades, Two Messages

In 1982 Mostafa Tolba, executive director of the UN Environment Program (UNEP), stated, "The world faces an ecological disaster as final as nuclear war within a couple of decades unless governments act now." He predicted that if the world failed to act immediately, there would be "by the turn of the century, an environmental catastrophe which will witness devastation as complete, as irreversible as any nuclear holocaust."[5]

In a 1988 interview James Hansen of NASA's Goddard Institute for Space Studies (GISS) was asked how he thought the area around his office in New York would be affected in 20 years. (2008).

James Hansen

Hansen replied: "The West Side Highway [which runs along the Hudson River] will be under water. And there will be tape across the windows across the street because of high winds. And the same birds won't be there. The trees in the median strip will change...There will be more police cars. . .[since] you know what happens to crime when the heat goes up." [6]

In 2008 he predicted that within five to ten years the Arctic would be free of summer ice. In a 2016 paper he stated: "The current rate of global warming could raise sea levels by 'several meters'[7] (21 feet) over the coming century, rendering most of the world's coastal cities uninhabitable and helping unleash devastating storms. . .Global warming of 2°C above pre-industrial times — the world is already halfway to this mark — would be "dangerous" and risk submerging cities."[8]

In 1990 Michael Oppenheimer of The Environmental Defense Fund made the following ominous prediction: "By 1995, the greenhouse effect would be desolating the heartlands of North America and Eurasia with horrific drought, causing crop failures and food riots. . ."[By 1996] "The Platte River of Nebraska will be dry, while a continent-wide black blizzard of prairie topsoil will stop traffic on interstates, strip paint from houses and shut down computers. . .The Mexican police will round up illegal American migrants surging into Mexico seeking work as field hands."[9] None of this ever happened, of course.

In 1990 Meryl Streep hosted a PBS 10-episode documentary entitled, "Race to Save the Planet". This documentary series predicted that by 2000 the world's average temperature would be four degrees warmer.

In January of 2000 Dr. Michael Oppenheimer of the Environmental Defense Fund, commenting in a *New York Times* interview on the mild winters in New York City, exclaimed: "But it does not take a scientist to size up the effects of snowless winters on the children too young to remember the record-setting blizzards of 1996. For them, the pleasures of sledding and snowball fights are as out-of-date as hoop-rolling, and the delight of a snow day off from school is unknown."[10]

The Big 2017 New York Snow

But that isn't what happened. New York City experienced several significant snowfalls during the winter of 2017-2018. And on March 12, 2018, a blizzard with gale winds up to 50 mph hit eastern North Carolina, several hundred miles south of New York. Parts of Kentucky received about a foot of snow in that same weather system.

On March 29, 2001, *CNN* stated: "In ten year's time [2011], most of the low-lying atolls surrounding Tuvalu's nine islands in the South Pacific Ocean will be submerged under water as global warming rises sea levels." Yet on Feb. 9, 2018, 17 years later, *The Daily Mail* reported that the land area of this chain of islands, comprised of coral atolls, is actually **growing**. A study conducted by the University of Auckland in New Zealand used aerial photographs as well as satellite imagery to demonstrate that eight of the atolls and nearly three fourths of the islands grew in size between 1971 and 2014.[11]In 2005, Andrew Simms, policy director of the New Economics Foundation, announced, "Scholars are predicting that 50 million people worldwide will be displaced **by 2010** because of rising sea levels, desertification, dried up aquifers, weather-induced flooding and other serious environmental changes." Fifty million people! This simply did not happen.

A February, 2004, headline in *the Guardian* reads, "A secret report. . . warns that major European cities will be sunk beneath raging seas as Britain is plunged into a 'Siberian' climate by 2020."

In June, 2008, Ted Alvarez of *Backpacker Magazine Blogs* wrote: "You could potentially sail, kayak, or even swim to the North Pole by the end of the summer.

Climate scientists say that the Arctic ice. . .is currently on track to melt sometime in 2008."[12] Later in that same year a Russian icebreaker was stranded for a week in the ice of the Northwest Passage.[13] The truth is that Arctic ice has been at approximately the same level for the past thirty years, and we have the data to prove it.

The 1970s cooling scare and the present warming scenario share these same three conclusions:

1) Mankind's activities are responsible.

2) We thus have the ability to control the climate.

3) We of the industrialized world must atone for our sins.

The fact that none of these predictions and tipping points has ever come about is one reason we know that the anthropogenic (human-caused) global warming hypothesis is incorrect. After all, if a hypothesis or theory is correct, scientists should be able to make accurate predictions based on it. Astronomers can predict solar and lunar eclipses to the exact minute, because they understand the positions and motions of the sun, moon and planets.

So, for all of these reasons, we don't need to take the Climate Clock very seriously except to regard it as a supreme example of the folly of mankind and the extent to which the elites will go in their efforts to impose totalitarianism on the world's people.

---

1. Javier Vinos, "The 2023 Climate Event Revealed the Greatest Failure of Climate Science," Clintel, Dec. 31, 2025.
2. www.aim.org/aim-column/washington-post-gives-megaphone-to-far-left-group s-
3. www.aim.org/aim-column/washington-post-gives-megaphone-to-far-left-group s-
4. "Cold Facts About Pollution," The Southeast Missourian, Jan. 29, 1970
5. Engdahl, op. cit
6. https://stevengoddard.wordpress.com/2013/03/10/another-hansen-fail/
7. "Several" could be five or more meters. Several meters equals about 17 feet, or from the ground to the roof of a house.
8. www.theguardian.com/science/2016/mar/22/sea-level-rise-james-hansen-climat e-change-scientist
9. http://www.foxnews.com/tech/2010/12/30/botched-environmental-forecasts.ht ml
10. http://www.climatedepot.com/2014/02/18/flashback-2000-new-york-times-sled ding-and-snowball-fights-are-as-out-of-date-as-hoop-rolling/
11. http://www.dailymail.co.uk/sciencetech/article-5371535/Sinking-Pacific-nation -getting-bigger-study.html

12. https://thebigfeedblog.blogspot.com/2008/12/the-worst-climate-predictions-of-2008.html

13. http://losingamerica.truth-in-politics.info/?p=3142

Chapter 2

The Beginning of the Global Warming Lie

"We are on the verge of a global transformation. All we need is the right major crisis." —David Rockefeller, Club of Rome Executive Manager

Acid Rain and CFCs

During the 1970s and 80s there was a big, media-fueled acid rain scare. It was claimed that emissions of sulfur dioxide (SO_2) and nitrous oxide gas emitted from electric utility companies in the Midwest United States were damaging forests and lakes in the eastern US, since air masses there move from west to east. Damage to trees from acid rain and other environmental pollutants in the Black Forest of Germany and in Scandinavia had also been observed. Nitrogen oxides (NO_x) and sulfur dioxide (SO_2) react with water in the atmosphere to form nitric acid and sulfuric acid.

In 1980 the US government set up a 10-year national Acid Precipitation Assessment Program, costing half a billion dollars, to investigate this problem. Alarm over acid rain that was spread by the media allowed governments to pass regulations governing the sale of sulfur dioxide permits. Realizing how environmental concerns amplified by the media could generate big bucks, Senators John Heinz and Timothy Wirth cosponsored a bill known as Project 88. This legislation was geared to create a whole new market for buying and selling sulfur dioxide permits through "cap and trade" and making a boatload of money for those positioned to profit from it. Project 88 morphed into the Clean Air Act of 1990.

Although the media discussed this nonstop, these acid rain fears proved to be largely unjustified. Just one species of tree, found at a high altitude, suffered any substantial damage. The acidity of the water in lakes was later attributed to natural causes. The scientists who studied this problem concluded that it was more complicated than previously believed. Wide acceptance of the Montreal Protocol, banning the use of some chlorofluorocarbons (CFCs), that we were told caused a hole in the ozone layer, encouraged environmental activists to call for similar controls on carbon dioxide emitted by the combustion of "fossil fuels", without regard for the fact that carbon dioxide is not a pollutant.

Dr. Roger Revelle| alchetron.com

A prominent scientist and head of the Scripps Institute of Oceanography, Dr. Roger Revelle established the International Geophysical Year (IGY) in 1957, during which measurements of carbon dioxide began. Dr. Revelle wrote the first modern scientific paper linking human carbon dioxide emissions to rising temperatures. In 1968 Al Gore, who later became a US senator and then vice president, happened to take one of Dr. Revelle's courses at Harvard University. Although receiving a "D" in the course, Gore was so impressed by Revelle's hypothesis on global warming that he made a highly exaggerated version of it the topic of his 1992 book, *Earth in the Balance.*

In *Earth in the Balance* Gore referred to Dr. Revelle as his mentor and inspiration to combat the "crisis" caused by carbon dioxide that was emitted by humans. Gore claimed that the increase in carbon dioxide caused sea level rise to accelerate and global temperatures to rise.

Al Gore | hd-chart.com

He contended that plant and animal species were disappearing at an alarming rate. He referred to carbon dioxide as "pollution" and argued that the problems he predicted would result from overproduction and consumption of resources by mankind. Gore pointed to a growing global population that was depleting resources. Through his position on the Commerce Committee where he teamed up with John Kerry, who later became President Biden's Climate Czar, he politicized the global warming issue .

In the 1980s the global warming issue gained attention through Senator Al Gore's 1988 congressional hearing on the matter. The scheduling and staging of this hearing had been carefully planned. Senator Tim Wirth later bragged in a PBS *Frontline* interview: "We called the weather bureau and found out what was historically the hottest day of the summer...So we scheduled the hearing that day and Bingo, it was the hottest day on record in Washington, or close to it. . .We went in the night before and opened all the windows so that the air conditioning wasn't working inside the room."

The star witness of the hearing, James Hansen of NASA's Goddard Institute for Space Studies (GISS), a climate activist who had participated in protests, gave a memorable performance. Senator Wirth also revealed during a 2007 interview on *Frontline* : "I don't remember exactly where the data came from, but we knew there was this scientist at NASA who had really identified the human impact before anyone else had done so and was very certain about it. So we called him up and asked him if he would testify." During this hearing Hansen claimed that he was 99% sure that the earth was warmer than at any time in recorded history. Of course that was a lie, since the decade of the 1930s dust bowl days had been much warmer.

Hansen was rewarded for this service by being appointed Director of NASA's Goddard Institute of Space Studies (GISS), a prestigious title. He took advantage of this position by pushing the anthropogenic global warming (AGW) narrative. He also violated the Hatch Act, which had been passed in 1939, to limit political activity by bureaucrats. He participated in several climate protests, even one held outside of the White House.[1]

Dr. Revelle was alarmed about this Senate hearing, which had caused the issue to explode. He believed that Senators Gore and Wirth were going too far. They took as facts many assumptions of which he was uncertain. So, shortly after this hearing, in July of 1988, he wrote a letter to Senator Wirth, warning, "We should be careful not to arouse too much alarm until the rate and amount of warming becomes clearer. It is not yet obvious that this summer's hot weather and drought are the result of global climatic change or simply an example of the uncertainties of climate variability. My own feeling is that we had better wait another ten years before making confident predictions." Senator Wirth ignored his letter.

Dr. S. Fred Singer

In 1991 Dr. Revelle, along with Drs. Fred Singer and Chauncey Starr, wrote an article entitled, "What to Do About Greenhouse Warming: Look Before You Leap". In this article they expressed doubts about the human-caused global warming hypothesis. They warned that more research needed to be done before taking the kind of drastic action that Senator Gore and others were calling for.

In their article, "Look Before You Leap," the three scientists stated, "The scientific base for greenhouse warming includes some facts, lots of uncertainty and just plain lack of knowledge, requiring more observations, better theories, and more extensive calculations. . .There is uncertainty about the strength of sources and sinks for these gases, i.e. their rates of generation and removal. There is major uncertainty and disagreement about whether this increase has caused a change in the climate during the last century."

But in his book *Earth in the Balance* Gore described Roger Revelle's views as much more alarmist than the concepts presented in this article. Gore wrote, "The implications of his [Dr. Revelle's] words were startling. We were looking at only eight years of information, but if this trend continued, human civilization would be forcing a profound and disruptive change in the entire global climate."[2]

According to John Coleman, founder of the Weather Channel, in 1992 Dr. Revelle delivered a speech at Bohemian Grove, a legendary club for elite gentlemen in California, in which he apologized for inadvertently leading so many in the wrong direction. He expressed his worries about the fear mongering that Al Gore and the UN Intergovernmental Panel on Climate Change (IPCC) were doing. He knew the science was flawed. Just a short time later, on July 15, 1992, he died of a heart attack.

On July 6, 1992, an article by Gregg Easterbrook, of *Newsweek* magazine, discussed "Look Before You Leap," the article that Drs. Revelle, Singer and Starr had written. He pointed out that in his book, *Earth in the Balance,* Gore claimed that Dr. Revelle, as his mentor at Harvard, had introduced him to the problem of global warming. "But," he noted, "*Earth in the Balance* does not mention that before his death last year, Revelle published a paper that concludes, 'The scientific base for greenhouse warming is too uncertain to justify drastic action at this time. There is little risk in delaying policy responses.'"

When the discrepancy between Gore's book and Revelle's article was raised during the 1992 Vice Presidential debate, Gore responded that Revelle's comments had been taken completely "out of context".

Shortly thereafter, another of Dr. Revelle's students, Dr. Justin Lancaster, called Dr. Singer and demanded that he remove Dr. Revelle's name from the article. Dr. Singer responded that this would be impossible, because he would need Dr. Revelle's permission to do so. And since Dr. Revelle had died, he would not be able to give it.

So Dr. Lancaster and Anthony Socci, a member of Gore's staff, spread the false story that Dr. Singer had added Dr. Revelle's name to the article to give it credibility. They claimed that Dr. Revelle was too feeble and senile to have written it. To counter these false and damaging statements Dr. Singer had no other option but to sue Dr. Lancaster for libel.

In the lawsuit's discovery process Dr. Singer learned that Lancaster had been fully aware of Dr. Revelle's contribution to the article. From the beginning of this controversy he knew that Dr. Singer had the galley proofs with Dr. Revelle's handwritten notes in the margins demonstrating his active participation in the article's preparation. Also, Dr. Revelle's calendar showed a full schedule of speaking engagements, conferences, and travel in the months before his death. This, of course, indicated that Dr. Revelle was not a cognitively impaired invalid as Lancaster and Socci had claimed. The court also found a lot of evidence linking Dr. Lancaster to Senator Gore.

In February of 1994 journalist Ted Koppel, on his TV program "Nightline", revealed that Al Gore, who now was Vice President, had called and asked him to investigate the political and economic forces behind the "antienvironmental movement", especially that Dr. Singer and other global warming skeptics had received money from the coal industry.

Ted Koppel asked Gore, "Is this a case of industry supporting scientists who happen to hold sympathetic views or scientists adapting their views to accomodate industry?"

Ted Koppel

Ted Koppel continued, "There is some irony in the fact that Vice President Al Gore, one of the most scientifically literate men in the White House in this century – is resorting to political means to achieve what should ultimately be resolved on a purely scientific basis. The measure of good science is neither the politics of the scientist nor the people with whom the scientist associates. It is the immersion of hypotheses into the acid of truth. That's the hard way to do it, but it the only way that works."

This attempt to use Mr. Koppel to vilify skeptical scientists brought criticism down on Al Gore. Lancaster's attorneys indicated they were ready to have him sign a retraction and apology for the lies he had told about Dr. Singer.

In his retraction and apology Dr. Lancaster testified to the court, "I agree to not make any such statements in future, and I apologize to Professor Singer for the pain my conduct has caused him and for any damage I may have caused to his reputation."

But after the lawsuit had been concluded, despite the sworn statement he had made to the court "to not make any such statements in future," Lancaster started repeating the same charges. He was thus admitting that he had lied under oath. In his retraction he had admitted that Professor Revelle was a true author of the paper. But then, contrary to these words, he insisted, "Revelle did not write it." He retracted his retraction despite documentary evidence in Revelle's own handwriting and numerous testimonials from others that Revelle did contribute to the article.

"To me it is astounding. It's absolutely astounding that we are having the living trying to take the names of the dead off papers that they wrote because the living don't agree with what the dead said when they authored the paper. That's straight out of a rather frightening book by George Orwell," exclaimed Dr. Pat Michaels, another global warming skeptic.

And of course they did this because Dr. Revelle, a highly respected scientist who was credited as the source of concern about the climate, was speaking out against Gore's agenda.

But now, as you might expect, all of this has been covered up. Google and other allies of the UN and the World Economic Forum (WEF) want you to believe the opposite of what actually happened. When you put the key words "Revelle, Gore, Singer" into a Google search, many articles such as the one from OSS (Open Source Systems) surface.

This website contends, "The 1991 *Cosmos* paper, 'What to Do About Greenhouse Warming: Look Before You Leap', is not 'Revelle's article'. That is a myth perpetuated at first by a few participants and now being spread by the many people they have misled. Revelle did not write it. The *Cosmos* article is S. Fred Singer's work. Singer may urge that Revelle was a co-author technically, simply because he allowed his name to be used."

These words are presented as fact, whereas the opposite is actually true, as evidenced by a court of law that investigated the entire affair and which had concluded that the men who claimed otherwise were lying.

A statement in Dr. Revelle's 1988 letter to Senator Wirth is nearly identical to one made on the first page of "Look Before You Leap". In that letter Dr. Revelle told Senator Wirth, "My own feeling is that we had better wait another ten years before making confident predictions." (emphasis added)

"Look Before You Leap" states, "There is little risk in delaying policy responses to this century old problem since there is every expectation that scientific understanding will be substantially improved within the next decade." (emphasis added)

Other online sources that deny the truth about this are: the Yale Review, Naomi Oreskes, Wikipedia, Pomona College and DeSmog Blog. In 2006 Naomi Oreskes and Erik Conway wrote a book entitled, *Merchants of Doubt*, in which they vilified scientists who, they said, cast doubt on "well-established scientific findings". They obviously didn't understand that science without skepticism isn't science.

They highlighted the controversy between Justin Lancaster and Fred Singer, falsely asserting, "Singer actively worked to discredit Revelle's work." They claimed that Revelle's papers clearly demonstrated that he had never changed his mind about global warming. To fact check these statements we need to examine what Dr. Revelle was thinking and saying in the years before that article was written.

In the November, 1982, *Scientific American* letters to the editors (nine years before "Look Before You Leap") Revelle wrote, "We must conclude that until a warming trend that exceeds the noise level of natural climatic fluctuations becomes clearly evident, there will be considerable uncertainty and a diversity of opinions about the amplitude of climatic effects of increased CO_2 . If the modelers are correct, such a signal should be detectable within the next 10 or 15 years."

Open Source Systems asked, "Did S. Fred Singer trick Dr. Revelle into an association to achieve an agenda? When the sequence of events is examined, the intentions seem to implicate his agenda."

However, the truth is that Drs. Singer and Revelle were longtime professional associates. They were jointly involved in several projects, among them a 1968 symposium under the auspices of the American Association for the Advancement of Science (AAAS) on the global effects of environmental pollution. They were both science advisors to the Department of the Interior. In the 1970s Dr. Revelle invited Dr. Singer to spend a few months as his guest at the Center for Population Studies at Harvard where he was a visiting professor. During that time Dr. Singer even stayed at Dr. Revelle's home for a few weeks.

Dr. Singer later served on an AAAS committee chaired by Dr. Revelle. After Revelle's return to California Dr. Singer would visit him from time to time at the Scripps Institution. At the AAAS convention in February of 1990 they met at breakfast and jointly decided to write the article "Look Before You Leap".

Here is a summary of a speech delivered by Dr. Revelle in 1980 — 11 years before "Look Before You Leap" was published. It is still available on YouTube:

1. Carbon dioxide is the world's most important chemical compound.
2. An increase in the percentage of carbon dioxide could cause more low clouds, increasing the albedo (reflectivity of the sun's rays), and thus cooling the planet.
3. The Medieval Warm Period — around 1000 AD — was warmer than the present.
4. The warming from the years 1910 to 1940 was not caused by carbon dioxide emissions. During the 1940 – 1980 cool down the percentage of carbon dioxide was rising.
5. Computer models are grossly inaccurate, because they don't take the existence of the oceans in any realistic way. "They treat the ocean as a swamp."
6. Increased photosynthesis from the rise in the percentage of carbon dioxide would boost agricultural yields and benefit everyone.
7. Much more research was needed. More would be understood in 10-15 years.

Does this sound like the points made in *Earth in the Balance*, which Gore claimed were inspired by Dr. Revelle?

Here is part of an interview of Dr. Revelle by *Omni Magazine* in March, 1984, seven years before the contested article:

OMNI: "A part of a problem that has occupied your attention for many years is the increasing level of CO_2 in the atmosphere, which could cause the Earth's climate to become warmer. Is this actually happening?"

REVELLE: "I estimate that the total increase [in CO_2] in the past hundred years has been about 21%. But whether the increase will lead to a significant increase in global temperature we can't absolutely say."

OMNI: "[If it happens], what will the warming of the earth mean to us?"

REVELLE: "There may be lots of effects. Increased CO_2 in the air acts like a fertilizer for plants. . .you get more plant growth. Increasing CO_2 levels also affects water transpiration, causing plants to close their pores and sweat less. This means plants will be able to grow in drier climates."

OMNI: "Does the increase in CO_2 have anything to do with people saying the weather is getting worse?"

REVELLE: "People are always saying the weather is getting worse. Actually the CO_2 increase is predicted to temper weather extremes."

Al Gore, who is not a scientist, has received monumental acclaim, including two Oscars, a Nobel prize, and millions of dollars for his books and his 2006 documentary, *An Inconvenient Truth*, which is riddled with many serious errors. In 2007 he told the House Energy and Science committee, "As soon as energy has a price, you're going to see a wave [of investment] in it." The company he started, Generation Management, which deals in cap and trade, has helped raise his net worth from $1.7 million to over $300 million.

Enter Enron Corporation

You may have heard of the Enron scandal, one of the biggest bankruptcy cases in US history. In the 1990s Enron held over $60 billion in assets and was doing a billion dollars in business per week. Due to fraud this company collapsed in 2001, along with Arthur Anderson LLP, one of the largest auditing and accounting companies in the world.

This event has had long-lasting repercussions. On July 7, 2004, Kenneth Lay, the CEO of Enron, was indicted by a federal grand jury for his involvement in the scandal. Many other Enron executives were indicted on a variety of charges and later sentenced to prison.

In the early 1990s Enron was the major trader in the $20 billion-per-year sulfur dioxide cap and trade market. The profits it made on these trades caused Enron's stock to rise significantly. Then came the inevitable question: "What next? Well, how about a carbon dioxide cap and trade program?" Enron recognized this as a dream opportunity. After all, since a cap and trade market for sulfur dioxide was already in place, why not do the same for CO_2, which was being blamed for the "climate crisis"?

Enron operated a vast network of natural gas pipelines. Because coal fired power plants emit more CO_2 than natural gas, its CEO, Ken Lay, saw the opportunity to make immense profits on carbon trading with the coal companies. Coal burning utilities would have to pay billions for permits, giving Enron a competitive advantage. Enron and other natural gas companies would make a fantastic amount of money selling permits – as well as their own energy – at higher prices.

Enron executives began working closely with Senator Gore, and they even hired James Hansen of NASA as a consultant. The Heinz Foundation, headed by Theresa Heinz Kerry, John Kerry's wife, awarded him $250,000.

But unless CO_2 was designated as a pollutant, its emissions couldn't be capped. As the newly-elected vice president in the Clinton administration in 1993, Al Gore soon latched on to the idea of a global environmental regulatory system. At the same time, Enron lobbied President Clinton and Congress to give the Environmental Protection Agency (EPA) the authority to declare carbon dioxide a pollutant.[3]

Ken Lay worked closely with Wirth, who now was Undersecretary of State for Global Affairs in the Clinton-Gore administration, to lobby Congress to grant EPA the regulatory authority to declare CO_2 a pollutant.

From 1994 to 1996, the Enron Foundation gave almost a million dollars to the Nature Conservancy, which was preaching global warming alarmism. Enron also bestowed nearly $1.5 million on other environmental groups that were publicizing global warming and developed close ties with groups like Greenpeace. At the same time its directors were working closely with the Clinton administration to push global warming propaganda.

Enron then bought the largest wind power company in the world, GE Wind, from General Electric. They also bought the world's largest solar power company. They gave research grants to scientists with the understanding that their reports would assign blame (attribution) for global warming to human carbon dioxide emissions.

In 1997, before the Kyoto protocol was adopted, an Enron attorney, Christopher Horner, was appointed Director of Relations with the federal government. Horner reported that on the second day at this job he was told that Enron's prime objective was to obtain an international treaty imposing cuts in CO_2 emissions. An internal Enron memorandum stated that Kyoto would "do more to promote Enron's business than almost any other regulatory initiative outside the restructuring [of] the energy and natural gas industries in Europe and the United States."[4]

The Kyoto Protocol

It was the Kyoto Protocol, passed on Dec. 11, 1997, that established the cap and trade markets for carbon dioxide. And it was Enron that pushed hardest for this agreement for its own benefit. Due to propaganda and pressure, governments went along with this. It set binding emission targets for 37 industrialized countries and economies in transition and the European Union – called Annex B countries.

On Sept. 1, 1998, Ken Lay, Enron's CEO, wrote a letter to President Clinton, asking him to shut down public debate on the matter and establish a Blue Ribbon Commission that, among other things, would vilify scientists who disagreed with the IPCC.

Al Gore and his partner David Blood, the former chief of Goldman Sachs Asset Management (GSAM), founded the Chicago Climate Exchange (CCX), which was set up to make record profits selling CO2 offsets when a cap and trade bill would be passed by Congress. CCX ceased trading in 2010 due to inactivity in the US cap and trade market.

Although "social justice" and "the environment" are their mantras for grabbing the moral high ground, those in power have denied loans to African countries such as Nigeria to install desperately-needed coal-fired power plants. If they use coal to produce electricity, Nigeria's thirty million people will lose the aid they receive from the European Union. These nations must promise to install only "alternative" energy that is impractical and 2-5 times as expensive.

As a result, the over 1.2 billion people who still don't have electricity must pay a high price for it. Only the rich people in those countries can afford that luxury. The poor are blocked from using cheap, abundant energy to pull themselves out of poverty. The price of food has skyrocketed, partly because some food crops, such as corn and sugar cane, have been diverted to the production of ethanol and other biofuels. This misappropriation of agricultural land caused food riots in Egypt, directly causing "Arab Spring," a wave of antigovernment protests in Egypt, Tunisia and other countries in 2010 and 2011. Is this social justice?

The well-placed elite have also instituted carbon trading, a practice yielding billions of dollars. In this scheme, also called emissions trading, or "cap and trade", businesses are forced by governments to trade their "pollution rights" through a complicated system of regulations. The companies producing less CO_2 can sell their unused "pollution rights" to companies that produce more carbon dioxide. This, of course, provides a financial incentive to produce less CO_2. But it is the consumers and taxpayers who ultimately pay the bill for carbon trading in the form of higher prices and taxes.

By the year 2008 $126 billion had already been paid out in these cap-and-trade schemes. This practice alone is expected to generate 2 to 10 trillion dollars for those set to benefit. Just one cap and trade auction held in California in February, 2022, generated about $975 million in revenue.[5] The cap and trade scheme was engineered by the bankers of such firms as J P Morgan Stanley, Barclays, Citigroup and Credit Suisse, who, in brokering these deals, are siphoning money away from hard working consumers and taxpayers. Casey Mulligen, reporting in the *New York Times*, stated that in the US "Some estimates suggest that the amount at stake for the taxpayers is [annually] over $3,000 per person."[6]

So the hypothesis of human-caused climate change has reaped huge profits for bankers, solar and wind corporations and companies set up to benefit from cap and trade. It has also provided careers for a new generation of scientists with a vested interest in this "problem". It has become obvious to many that the climate change agenda is not about climate change. It is about windfall profits and total control of the people.

---

1. Dr. Tim Ball, *Human Caused Global Warming: the Greatest Deception in History*, 2016, p.20.
2. Al Gore, *Earth in the Balance*, 1992, p. 5
3. Bell, Larry, *Climate of Corruption*, Greenleaf Book Group Press, Austin, TX, 2011, p. 129
4. Bell, Op Cit, p. 137
5. https://lao.ca.gov/Publications/Report/4591
6. https://economix.blogs.nytimes.com/2009/05/27/how-much-will-cap-and-trade-cost/

Chapter 3

Questionable Motives

"If I were reincarnated, I would wish to be returned to earth as a killer virus to lower human population levels." —Prince Philip, Duke of Edinburgh and patron of the World Wildlife Fund

The Overpopulation Hypothesis

A Canadian oilman who made a huge fortune in the US, Maurice Strong played an important role in establishing the Intergovernmental Panel on Climate Change (IPCC) in 1988. A friend of the oil magnate David Rockefeller, Strong was appointed chairman of the 1992 Earth Summit United Nations Conference, at which he advocated population reduction and the lowering of living standards in the interest of "saving the environment." His radical viewpoints were revealed when he asked, **"Isn't the only hope for the planet that the industrialized civilizations collapse? Isn't it our responsibility to bring that about?"**[1]

Maurice Strong

Strong had helped found the Club of Rome in 1968. The members of this organization believed that the earth was overpopulated and that its resources were being used up too fast, especially by the industrialized nations. Many of their beliefs stemmed from the mistaken idea first advanced by Thomas Malthus in his 1798 *Essay on the Principle of Population* that the world's population would outgrow the food supply.

Thomas Malthus |teepu blic.com

Malthus' conclusions were disproved by what actually happened. As the world's population increased, the percentage of people who were thriving also increased.

Malthus' deluded philosophy contributed to the ideas of Charles Darwin[2] and thus the development of the eugenics movement in the early 1900s. This movement, based on warped thinking, advocated the selective breeding of the human population – by methods such as the sterilization of those deemed inferior – to improve its genetic composition. But who would decide who was superior and who was inferior? Obviously it would be those in power. This led to the unfortunate episode of the Holocaust carried out by the Nazis during World War II.

In his book, *The Population Bomb*, published in 1968, Paul Ehrlich and his wife, Anne, revived Malthus' idea. In this book the Ehrlichs predicted, "The battle to feed humanity is over. In the 1970s, the world will undergo famines. Hundreds of millions of people are going to starve to death in spite of any crash programs embarked upon now. Population control is the only answer."

None of Ehrlich's forecasts materialized. Yet there are still many people, even some billionaires[3], who believe the world's population should be as low as 500 million – not the current 8 billion. They believe that if Third-World people get the energy they need to raise their living standards, the earth's population will rise further.

Paul Ehrlich

Actually the opposite is true. History has shown **in every instance** that when a population's living standards increase, the birthrate decreases.[4] First the death rate comes down. In a few years a decrease in the birthrate occurs.[5] This principle is known as the Demographic Transition, as shown in this graph.

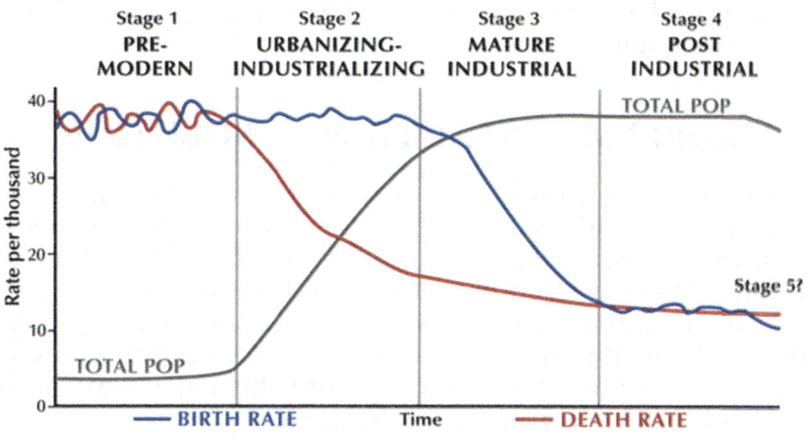

The Demographic Transition

History has repeatedly demonstrated that industrialized civilizations have lower birth and death rates than those of third-world nations. In asserting that the world's rapid population growth was unsustainable, Ehrlich proposed radical population control measures, including sterilization. In 1977 he and John Holdren, who later became Obama's science czar, wrote a book entitled, *Ecoscience: Population, Resources, Environment.* This book called for forced abortions, contraceptive drugs in water supplies, seizing babies from single and teen mothers, forced sterilization of those they deemed unfit and an international police force.[6] The concept of eugenics, which inspired the Holocaust perpetrated by the Nazis, has been around for a long time. In a 1970 editorial in *The Economist*, Ehrlich revealed his opinion of mankind as a cancer on the earth, stating,

> "The extinction of the human species may not only be inevitable but a good thing. . .This is not to say that the rise of human civilization is insignificant, but there is no way of showing that it will be much help to the world in the long run. . .Giving society cheap, abundant energy. . .would be the equivalent of giving an idiot child a machine gun."

In 1971 Ehrlich declared, "England will not exist in the year 2000." Not a single one of his many predictions has materialized. But not in his mind. In 2023, he bragged,

"60 Minutes extinction story has brought the usual right wing out in force. If I'm always wrong, so is science, since my work is always peer-reviewed, including THE POPULATION BOMB, and I've gotten virtually every scientific honor. Sure, I've made some mistakes, but no basic ones."

The Club of Rome's frightening goal and strategy based on these mistaken ideas was revealed in its book, *The First Global Revolution* (1991), which asserts,

> "In searching for a new enemy to unite us, we came up with the idea that pollution, the threat of global warming, water shortages, famine and the like would fit the bill. . .All these dangers are caused by human intervention. . .The real enemy, then, is humanity itself."[7]

Here they had hit on something for which to make people feel guilty. A well known strategy used by governments for grabbing power that has been known and used for a long time is:

PROBLEM → REACTION → SOLUTION

Those in power create a problem or crisis that is real or imagined. They sell this "problem" to the people using thousands of reports and articles — cleverly contrived propaganda. The public reacts by begging for a solution, which is then provided handily by the people in power. The rulers say, "We're here to solve the crisis (we created)." Because global warming is a worldwide "problem", the "solution" is too difficult for nation states to handle by themselves. Thus, the ultimate solution to this particular "problem" is nothing short of totalitarianism on a global scale.

"It is simply not feasible for sovereignty to be exercised unilaterally by individual nation-states, however powerful. It is a principle which will yield only slowly and reluctantly to the imperatives of global environmental cooperation." —Maurice Strong

> "In nature organic growth proceeds according to a Master Plan, a blueprint. Such a "master plan" is missing from the process of growth and development of the world system. Now is the time to draw up a master plan for sustainable growth and world development based on global allocation of all resources and a new global economic system."[8]

Just as in the book, *1984*, a common enemy and a common cause around which the entire planet could rally had been found. Environmentalism, an important and noble cause, was hijacked by the ruling elite who have no interest in the environment, as demonstrated by their huge mansions and private jets in which they regularly fly to exotic places for climate conferences.

Under Strong's direction as Secretary-General of the UN Conference on the Human Environment, the first Earth Day was held in April, 1970. In 1972 Strong founded the UN Environment Programme (UNEP). In the same year the UN Stockholm Conference was held in Sweden. This led to the Stockholm Declaration, which laid out a number of principles on environmentalism.

In 1979 the first World Climate Conference was held at Geneva, Switzerland. This was sponsored by the UN World Meteorological Organization (WMO) and the International Council for Science (ISCU). At the forefront of their agenda was a paper by Professor Bert Bolin, which established a connection between human-generated carbon dioxide and global warming. Maurice Strong realized this would help him achieve his goal of a world government by the United Nations. [9]

The first International Conference on Climate Change was held in Villach, Austria, in 1985. This was followed in 1987 by the Brundtland Report, "Our Common Future", which defined sustainable development as meeting the needs of the present without compromising the ability of future generations to meet their own needs.

The UN Intergovernmental Panel on Climate Change (IPCC) was founded by the UNEP and the World Meteorological Organization (WMO). Strong was an organizational genius. The charter of this organization reads,

> "The role of the IPCC is to assess on a comprehensive objective, open and transparent basis the scientific, technical and socioeconomic information relevant to understanding the scientific basis of risk of **human-induced climate change**, its potential impacts and options for adaptation and mitigation." (emphasis added)

One way we know that the IPCC is a political — not scientific — organization is that its charter does not call for finding the **actual causes** of global warming. Its creators simply **assumed** that it is caused by carbon dioxide emitted by humanity.[10]

In 1992 Strong organized the UN Framework Convention on Climate Change (UNFCCC) at the Rio Earth Summit. It was at this conference that the globalist-inspired UN Agenda 21 (now called "Agenda 2030") was also established.[11] After the first Earth Summit in Rio De Janeiro, Strong was appointed Assistant Secretary General of the United Nations.

The enforcement and decision making arm of the UN Framework Convention on Climate Change (UNFCCC) is the Conference of the Parties (COP), which holds conventions yearly in different locations. The first, COP-1, was held in Berlin in 1995. The purpose of COP is to implement the policies of the UNFCCC and its related agreements (Kyoto Protocol in 1997 and the Paris Agreement in 2015). The "parties" are about 200 nations which send representatives to these conferences. Participants negotiate and adopt new decisions and rules for climate mitigation. These are spectacular events, housing between 30,000 and 70,000 participants — who arrive in thousands of private jets — and sometimes place a real burden on the country or city entertaining them.

It's no accident that the UNFCCC and the Agenda 21 plan were given birth at the same time. The UNFCCC and IPCC would provide the excuse for the establishment of Agenda 21. According to the investigative journalist, Elaine Dewar, "Strong was using the UN as a platform to sell a global environment crisis and the global governance agenda."[12]

In her book, *Cloak of Green*, Dewar related a conversation she had with Strong. She asked him why he didn't run for office to achieve his goal of bringing about the collapse of industrialized nations. He responded that it would be much easier working at the UN. "He could raise his own money, from whomever he liked, appoint anyone he wanted, control the agenda."[13]

Dr. Henry Lamb, who founded the Climate Research Unit (CRU) in Britain, understood what Strong was doing. In 1997 he remarked, "Strong prefers to operate in the background. He, perhaps more than any single person, is responsible for the development of a global agenda now being implemented throughout the world." In his 1996 book, *The Rise of Global Governance*, Lamb explained, "Strong has worked diligently and effectively to bring his ideas to fruition. He is now in a position to implement them."

And of the United Nations he warned, "Education programs to teach the 'global ethic' have been underway by UNESCO [United Nations Educational Scientific and Cultural Organization] and by UNEP for more than 20 years. That the US government, through its representatives to the various UN agencies, has not already crushed this global governance agenda is a testament to the effectiveness of the UN's education program."

There would be no place in Strong's globalist government for either democracy or the ability to disagree openly. There would be no legal protection for whistle-blowers. There would be no veto power by any nation. As Lamb concluded, "The fox has been given the assignment, and all the tools necessary, to repair the hen house to his liking."

Agenda 21 is supposedly a voluntary action plan for sustainable development. It is an action plan to inventory and control **all** land, **all** water, **all** minerals, **all** plants, **all** animals, **all** construction, **all** means of production, **all** information, **all** energy and **all** human beings. In other words, it is pure socialist or fascist totalitarianism.

Strong set up the United Nations Environmental Program (UNEP) and the Intergovernmental Panel on Climate Change (IPCC) to create the science he needed to advance the Club of Rome's political agenda and to pressure politicians to take action. The IPCC was set up under the World Meteorological Organization (WMO) at the 1985 Villach, Austria, meeting. Climate scientists were appointed to the IPCC to research the subject and write reports which would lend the scientific authority for the policy reports to be handed to individual governments.[14]

The UNFCCC literally controlled the outcome of this research by narrowing its focus. It set the ground rules for climate science in order to predetermine the claim that carbon dioxide was causing the temperature increase. In their research the IPCC scientists were permitted to consider **only** the IPCC's definition of the term "climate change": "a change of climate **which is attributed directly or indirectly to human activity** that alters the composition of the global atmosphere and which is, in addition to natural climate variability, observed over considerable time periods." (emphasis added)

When the IPCC was set up, two stipulations were made. First, it was established that all Summaries for Policymakers were to be approved line by line by member governments, demonstrating that this report is just a government report with no scientific value. The second stipulation, following closely behind the first, is that Summaries for Policymakers override any differing conclusions scientists write for IPCC reports.

These scientists were not permitted to consider any other possibilities. The IPCC would write its Summary Report for Policymakers **before** the scientific report had even been submitted. They would then take that report to the scientists and tell them, "Make sure your report agrees with this." In their First Assessment Report, written in 1990, the IPCC scientists truthfully acknowledged that, although the climate was changing, there was no direct evidence that any recent changes had been caused by mankind. They concluded:

> "None of the studies cited above has shown clear evidence that we can attribute the observed [climate] changes to the specific cause of increases in greenhouse gases. While some of the pattern-base discussed here have claimed detection of a significant climate change, no study to date has positively attributed all or part of climate change to man-made causes. Any claims of positive detection and attribution of climate change are likely to remain controversial until

uncertainties in the natural variability of the climate system are reduced."

Despite this conclusion by the IPCC scientists, the final policy report summary, composed by Dr. Ben Santer reads:

> "There is evidence of an emerging pattern of climate response to forcing by greenhouse gases and sulfate aerosols...from the geographical, seasonal and vertical patterns of temperature change. . .**These results point toward a human influence on global climate.**" (emphasis added)

So Santer totally disregarded the conclusions that the other IPCC scientists had reached and essentially wrote his own report.

The Summary for Policymakers report read, "The balance of evidence suggests a discernible human influence on global climate."[15]

To be consistent with the Summary for Policymakers, the report the scientists had written was redone, and the Summary for Policymakers' statement inserted in its place. This invalidated what the scientists had actually concluded. Fifteen sections of Chapter 8 of the scientists' report were rewritten.

In 2006 Avery and Singer described what had actually happened:

> "Santer single-handedly reversed the 'climate science' of the whole IPCC report and with it the global warming political process! The 'discernable human influence' supposedly revealed by the IPCC has been cited thousands of times since in media around the world and has been the "stopper" in millions of debates among nonscientists."

Although most IPCC scientists are competent and truthful, few have had any influence over final IPCC conclusions. Their work is cherry picked line by line by political appointees, many of whom are not scientists at all but activists from environmental groups such as the World Wildlife Fund. These "Summaries for Policymakers" always end up saying what the IPCC wanted them to say all along and are then passed on to the media and the public. This caused Frederick Seitz, a world-famous physicist and former president of the US National Academy of Sciences, the American Physical Society and Rockefeller University, to write, "I have never witnessed a more disturbing corruption of the peer review process than events that led to this IPCC report."[16]

Other top scientists who had formerly supported the IPCC have protested its unscientific methodology. Professor Fritz Vahrenholt, a German environmentalist, also criticized the IPCC for its shady practices, incompetence and dishonesty in its gross exaggeration of the influence of human CO2 emissions on the climate. He wrote a book entitled, *The Cold Sun: Why the Climate Disaster Won't Happen.*

The IPCC fooled most everyone, even scientists. They didn't believe that science would be used to advance a political agenda. For example, Klaus-Eckart Puls remarked,

> "Ten years ago I simply parroted what the IPCC told us. One day I started checking the facts and data. First I started with a sense of doubt, but then I became outraged when I discovered that much of what the IPCC and the media were telling us was sheer nonsense and was not even supported by any scientific facts and measurements. To this day I still feel shame that as a scientist I made presentations of their science without first checking it. . . Scientifically it is sheer absurdity to think we can get a nice climate by turning a CO2 adjustment knob."[17]

In 1993 US Senator Tim Wirth (D-CO), who was on board with all of this, revealed the true motivation for this deception. "We've got to ride the global warming issue," he said. "Even if the theory of global warming is wrong, we will be doing the right thing."

Sometimes the perpetrators of a deception put the truth right out in plain sight. In 1998 Christine Stewart, Canadian Minister of the Environment, commented, "No matter if the science is all phony, there are collateral benefits. . .Climate change [provides] the greatest chance to bring about justice and equality in the world."[18]We have to ask why someone in her position would make a comment like that.

Despite this promise of "justice and equality" the fulfillment of the IPCC goals by means of the "Green New Deal" or a similar plan would actually have the opposite effect. The IPCC's stated aim is to redistribute wealth from the rich, developed nations to the poor, third-world nations. Looking more closely, this plan really transfers funds from the poor of the rich nations to the corrupt rulers of the third-world nations, many of whom personally profit at the expense of their citizens.

In introducing *Agenda 21,* Maurice Strong boldly declared, "Current lifestyles and consumption patterns of the affluent middle class — involving meat intake, use of fossil fuels, appliances, air-conditioning, and suburban housing — are not sustainable."

In 1992 *Agenda 21* slipped under the radar into the United States. Since it was supposedly a "voluntary" plan, the people were never consulted or even informed of this trickery. In 1993 the Clinton Administration wasted no time in establishing the "President's Council for Sustainable Development" to begin implementing some of its goals.[19] Congresswoman Nancy Pelosi (D-CA) introduced a resolution of support for the plan into Congress, calling it a "comprehensive blueprint for the reorganization of human society."[20]

Anyone who disagrees with the official climate change narrative is called a "climate denier" (after Holocaust denier) or, even worse, a "science" denier. Those who resort to this type of ad hominem attack are revealing their lack of understanding of what science is. Just as Tony Fauci asserted that he is the science, the climate scientists spouting the official narrative believe they are the science. Those who have dared to disagree that mankind is responsible for all of the warming we've seen since 1850 have lost their grants, and sometimes even their jobs. Many of their papers have been withheld from scientific journals and are carefully avoided by the media. So the public believes the story that "97% of all scientists agree that climate change is experiencing dangerous warming caused by human activities."

The Climate Change Strategy

1. Get yourself set up as a scientific authority (UN IPCC, UNEP, etc.)
2. Create a crisis. Have the media run with it. Then take the moral high ground by painting yourself as an environmentalist and dealing with the problem: vilifying acid rain, DDT, freon and other CFCs, carbon dioxide and methane.
3. Capture academia and the scientific institutions with buckets of government grant money.
4. Establish cancel culture, and censor and smear people who disagree with you.

Nowhere do you see the implementation of cancel culture to the degree that has occurred with the climate change issue. The pressure to be politically correct is especially rampant in our universities and colleges. Simply mentioning the subject could put a professor who doesn't have tenure in danger of being fired. This is occurring even in corporations. Attempting to speak on this subject to climate change religious fundamentalists can quickly brand someone as a right-wing conspiracy theorist.

The Climategate Scandal

The term "Climategate" refers to a controversial episode in climate science that unfolded in late 2009 following the unauthorized release of thousands of emails and documents from the Climate Research Unit (CRU) located at the University of East Anglia in the United Kingdom. On November 19, 2009, an unknown person released over 1,000 emails and 3,000 other documents onto the internet. The leaked communications spanned a period of over a decade (1996 to 2009) and included exchanges between CRU scientists as well as correspondence with other climate scientists, IPCC officials and journalists. These leaked communications sparked intense debate and scrutiny of the methods and ethics employed by some of the world's leading climate researchers.

Established in 1972, the CRU is a prominent institution in climate science, known for its work on global temperature records and its role in the Intergovernmental Panel on Climate Change (IPCC) process. Phil Jones, a prominent climate scientist and the head of the CRU at that time, was a central figure in the Climategate controversy.

Some of these emails indicated that climate scientists had manipulated or suppressed data to support their agenda on global warming. They discussed withholding data from climate skeptics, with Phil Jones stating, "I think we should delete any emails that mention [the data]. I've deleted mine...Can you do likewise?" (Jones, 2007)

Some emails showed that these scientists were excluding skeptical voices and promoting those supporting their agenda. In an email discussing papers written by climate skeptic Chris de Freitas, Phil Jones wrote, "I can't see either of these papers being in the next IPCC report. Kevin and I will keep them out somehow – even if we have to redefine what the peer review literature is." (Jones, 2003)

In March of 2003 Phil Jones wrote to his colleagues:

"A paper by astrophysicists Willie Soon and Sallie Baliunas was published by *Climate Research*, which concluded that 'the 20th century is probably neither the warmest nor a uniquely extreme climate period of the last millennium.'

"Tim Osborn has just come across this. Best to ignore probably, so don't let it spoil your day. I've not looked at it yet. It results from this journal having a number of editors. The responsible one for this is a well known skeptic in New Zealand. He has let a few papers through [by skeptics] Michaels and Gray in the past. I've had words with Hans von Storch about this, but got nowhere. . .

"Writing this I am becoming more convinced we should do something. . .I will be emailing the journal to tell them I'm having nothing more to do with it until they rid themselves of this troublesome editor. A Climatic Research Unit person is on the editorial board, but papers get dealt with by the editor assigned by Hans von Storch."

The response to this from Mike Mann was:

"The Soon and Baliunas paper couldn't have cleared a 'legitimate' peer review process anywhere. That leaves only one possibility — that the peer review at *Climate Research* has been hijacked by a few skeptics on the editorial board. And it isn't just De Freitas; unfortunately, I think this group also includes a member of my own department. . .The skeptics appear to have staged a 'coup' at *Climate Research.* (It was a mediocre journal to begin with , but now it's a mediocre journal with a definite purpose.)"

In another group of emails released in 2011 Mann plotted to repair the "damage" from the Soon-Baliunas study, being referred to by Mann as "two awful papers written by those clowns" and "an assault on the science of climate change."

Mike Mann continued, "So what do we do about this? I think we have to stop considering *Climate Research* as a legitimate peer reviewed journal. Perhaps we should encourage our colleagues in the climate research community to no longer submit to or cite papers in this journal." [21]In their 2010 book, *The Crutape Letters,* Steven Mosher and Thomas Fuller described the CRU scientists' activities:

> "[They] ruthlessly suppressed dissent by insuring that contrary papers were never published and that editors who didn't follow their party line were forced out of their positions. When Freedom of Information requests threatened to reveal their misbehavior, the emails showed them actively conspiring to delete emails to frustrate legitimate requests for information. Worst of all, one scientist threatened to delete climate data rather than turn it over, and that data is still missing."

In preventing scientists whose findings disagreed with theirs from being published, this small clique was not only damaging the careers of other scientists but also killing debate and preventing alternative points of view from finding their way to the United Nations decision makers.

Equally shocking, these men revealed the ruthless methods they used to silence anyone who dared to question their conclusions. They had pressured editors of scientific journals to block publication of studies contradicting their conclusions. A professional scientist's career success rests on his/her ability to get studies published in professional journals. They discussed ways in which they could pressure the editors of these journals not to publish studies that had reached conclusions differing from theirs. This was killing the careers of scientists who disagreed with them.

Mann wrote, "Perhaps we should encourage our colleagues in the climate research community to no longer submit to or cite papers in this journal?"

To this Jones responded, "I will be emailing the journal to tell them I'm having nothing more to do with it until they rid themselves of this troublesome editor. And when Otto Kinne, editor of *Climate Research,* defended Chris de Freitas by saying he had done a good job, Michael Mann commented, "It seems to me that this 'Kinne' character's words are disingenuous and probably supports what de Freitas is trying to do. It seems clear we probably have to go above him. I think that the community should, as Mike Hulme has previously suggested in this eventuality, terminate its involvement with this journal at all levels — reviewing, editing and submitting, and leave it to wither away into oblivion and disrepute."

They also insulted dissenting scientists. For example, they called Piers Corbyn, an immensely successful meteorologist, a prat (British for "useless idiot").

Jones demonstrated corruption of the peer review process in several emails, such as the following to Mike Mann. "You might think Keith and I have reviewed some of your papers, but we haven't. I've reviewed Ray's and Malcolm's – constructively, I hope, where I thought something could have been done better. I also know you've reviewed my paper with Gabi Hegerl very constructively. [22]

Peer reviews are supposed to be made anonymously. But the Climategate emails indicated that members of this clique agreed to peer review one another's papers in order to get as many of them published in scientific journals as possible.

Emails also revealed the CRU scientists had refused freedom of information (FOI) requests for data and methods, citing privacy concerns or claiming that the data did not exist. This raised questions about transparency and accountability in climate research.

Other quotes from emails indicated what was going on:

In a letter to Michael Mann, Phil Jones wrote, "We have to stop this 2000-year paleoclimate nonsense. These idiots in the geologic community are saying that the medieval warm period was as warm as the present. We can't have that — it makes the recent increase in CO2 look trivial, which it isn't."

In another email he remarked, "I've been told that IPCC is above national laws. When I've discussed this with IPCC, they have laughed out loud." This implies that the Intergovernmental Panel on Climate Change (IPCC) considers itself above national laws and accountability.

An email from Michael Mann to Phil Jones reveals, "We also have a data problem. We can't find the missing 20 years of data. If we can't find it, we'll have to make it up." [23]This quote implies that scientists were willing to fabricate data to support their conclusions.

Some emails also revealed close relationships between CRU scientists and journalists, with scientists providing journalists with "talking points" or helping to shape news articles. In one instance CRU scientists Tom Wigley provided a journalist with talking points to use in an article.

In one email Kevin Trenberth lamented, "The fact is that we can't account for the lack of warming at the moment, and it is a travesty that we can't." (Trenberth, 2009) If he was really worried that the earth was warming, shouldn't he have been elated that temperatures had leveled off?

In another email, CRU scientist Tom Wigley admitted to manipulating data, stating, "We should maybe do a paper just saying that when some people do this, wrongly, it doesn't matter." (Wigley, 2007)

These emails detailed discussions among CRU scientists about the Hockey Stick graph constructed by Michael Mann of Penn State University. Mann and his helpers hooked thermometer data onto proxy data in order to get the result they wanted, something that should never be done in a scientific study. This greatly exaggerated the threat they claimed existed from global warming over the previous century. In one email Mann discussed the graph, stating, "I've been trying to figure out where the heck the 20th century values could be coming from."

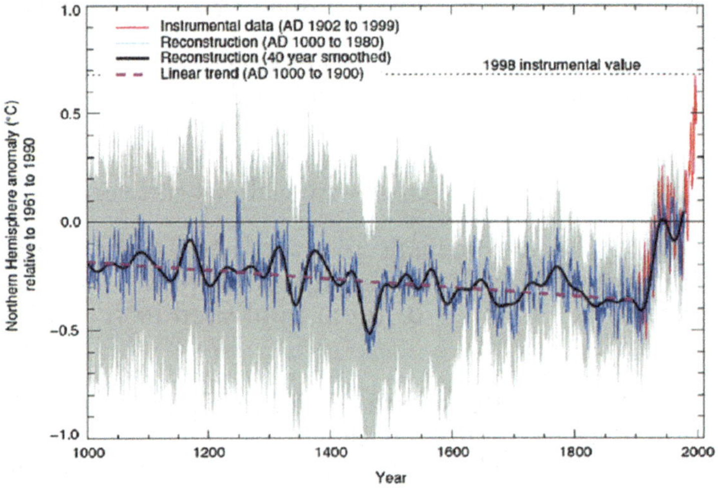

Michael Mann's "Hockey Stick" Graph

This graph totally obliterated the Medieval Warm Period and the Little Ice Age. It was published everywhere, including the prestigious *Nature Magazine*. But no one had bothered to check the data behind it.

The statistical methods Mann had used to create the uptick on the right side – which to some resembled a hockey stick – were first exposed as fundamentally flawed by Steve McIntyre, a Canadian statistician. A heated battle was raging between Mann's supporters and McIntyre over the statistical basis on which the CRU had constructed the global warming case for the IPCC. Based on data taken from bristlecone pine trees, the graph incorrectly showed that – after a thousand years of global temperatures being very steady – they had suddenly shot up to their highest level in recorded history.

The "hockey stick" graph was published everywhere, including the prestigious *Nature Magazine*. But no one had bothered to check the data behind it. "Given star billing by the Intergovernmental Panel on Climate Change, not least for the way it appeared to eliminate the long-accepted Medieval Warm Period, when temperatures were higher than they are today, the graph became the central icon of the entire man-made global warming movement."[24]

The tree ring data actually indicated that temperatures in the latter part of the 20th century had been declining. Because this wasn't what Mike Mann had anticipated, he decided to lop off the tree ring data from 1961 onward and replace it with data from thermometers. Compare Mann's Hockey Stick graph with the one constructed by the founder of the CRU, Hubert Lamb, and which appeared in the first IPCC report in 1990.

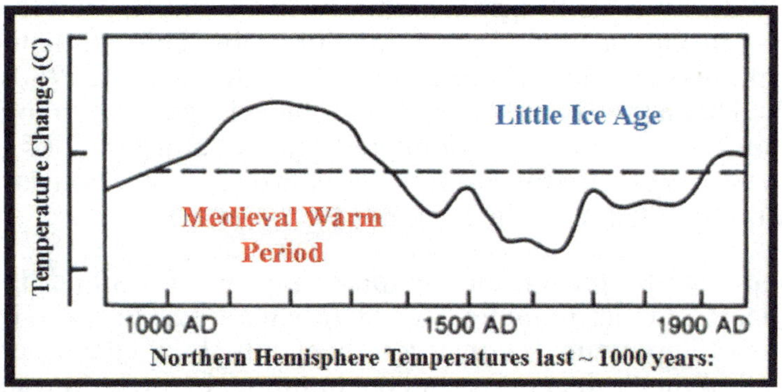

The leaked email documents also revealed discussions between these men about how they could eliminate data to get rid of the Medieval Warm Period, which lasted from about 950 AD to 1250 AD. Members of this elite group even considered tactics to prevent the release of the data on which the IPCC's conclusions were based, data requested by other scientists under Freedom of Information (FOI) laws, which incidentally is a criminal offense in Britain.

Phil Jones, Head of the Climate Research Unit, indicated his continued efforts to keep important data under wraps.

> "I've been told that IPCC is above national FOI acts. One way to cover yourself and all those working on AR5 would be to delete all emails at the end of the process. Any work we have done in the past is done on the back of the research grants we get – and has to be well hidden. I've discussed this with the main funder [US Dept. of Energy] in the past and they are happy about not releasing the original station data."

But if their research was being funded by the US Dept. of Energy, it was not proprietary data and should have been in the public domain.

Jones claimed to have lost large chunks of this data. He had commented, "I should warn you that some data we have we are not supposed to pass on to others. We can pass on the gridded data, which we do. Even if WMO agrees, I will still not pass on the data. We have 25 or so years invested in the work. Why should I make the data available to you, when your aim is to try and find something wrong with it?"[25]

He also indicated that there were serious credibility problems with the models they were using. "Basic problem is that all models are wrong – not got enough middle and low level clouds. . .What he [Zwiers] has done comes to a different conclusion than Caspar and Gene! I reckon this can be saved with careful wording."

This attitude violates the scientific method, which demands transparency. A defining characteristic of science is the sharing of data and procedures used in their study, so that independent parties can replicate and validate their observations and experiments. Scientists are expected to make requested data available to others who are endeavoring to understand how they reached their conclusions.

In the same year, 1998, Keith Briffa, another CRU scientist, produced a tree ring graph showing pronounced cooling in the mid 1900s. This conflicted with some ground thermometer readings. These men contended that the 1980s thermometer records had been lost. They solved this problem by making up some "modified data". Phil Jones wrote, "I've just completed Mike's nature trick of adding in the real temps to each series for the last twenty years (ie from 1981 onwards) and from 1961 for Keith's to hide the decline." (Jones 1999)

In a 1999 paper, Briffa presented a graph of proxy studies ending in 1961 and adding to it data from thermometers, which he called "real temps". Here is a graph of Keith Briffa's temperature reconstruction, plainly showing temperatures rising in the first half of the 20th century and declining again during the second half of the 1900s..

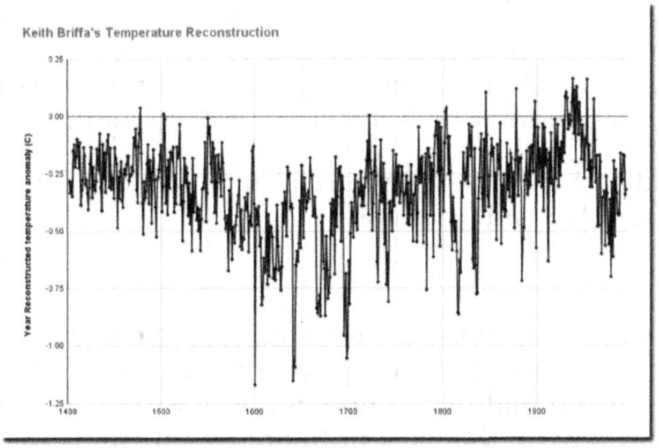

20th Century Temperature Record from Tree Rings

They demonstrated their hatred of skeptical scientists, rejoicing at the news that one of them had died. Ben Santer, the same scientist who had reversed the conclusions of IPCC scientists in their first report in 1996, told CRU, "The next time I see Pat Michaels [a dissenting scientist] at a scientific meeting, I'll be tempted to beat the crap out of him."

Several emails indicated their desire to make the results of their research as sensational as possible, whether correct or not. On Sept. 19, 1996, Gary Funkhouser confessed, "I really wish I could be more positive about the material, but I swear I pulled every trick out of my sleeve trying to milk something out of that...I don't think it'd be productive to try and juggle the chronology statistics any more than I already have – they just are what they are...I think I'll have to look for an option where I can let this little story go as it is."[26]

Geoff Jenkins' letter to Phil Jones on Nov. 22, 2009, reads in part, "Remember all the fun we had last year over 1995 global temperatures, with the early release of information (via Australia), 'inventing' the December monthly value, letters to Nature, etc., etc.? I think we should have a cunning plan about what we should do this year, simply to avoid a lot of wasted time...We feed this selectively to

Nick Nuttall (who has had this in the past and seems now to expect special treatment) so that he can write an article for the silly season...I know it sounds a bit cloak-and-dagger but it's just meant to save time in the long run."[27]

One scientist truthfully pointed out, "It is inconceivable that policymakers would be willing to make billion and trillion dollar decisions for adaptation to the projected regional climate change based on models that do not even describe and simulate the processes that are the building blocks of climate variability."[28]

You may have wondered how the term "global warming" changed to "climate change". This was no accident. A 2003 email from Minns/Tyndall Center on the University of East Anglia campus stated, "In my experience with global warming freezing is already a bit of a public relations problem with the media." To this came a reply from Bo Kjellen of Sweden, "I agree with Nick that climate change might be a better labeling than global warming."

A letter to the Norwich Green Party by Ian Harris, who worked for CRU, reveals the intense desire these scientists had to use scare tactics to force public policy: "We're looking at an unprecedented acceleration in temperature. . .Even if it turns out to be naturally-occurring, who's willing to take that chance? We should be trying to wean ourselves off of unsustainable energy generation and use anyway." Harris was saying in effect that even if the warming should later be found to be from natural causes rather than from the actions of mankind, he didn't care.[29]

In November of 2011 about 5,000 additional emails were leaked, some even more damning than the ones released previously. Scientists in this cabal were advised to delete large chunks of data. Doing this subsequent to an FOI request is a criminal offense in Britain.

Myron Ebell commented, "If there were any doubts remaining after reading the first Climategate emails, the new batch of emails that appeared on the web today [November 22] make it clear that the UN Intergovernmental Panel on Climate Change is an organized conspiracy dedicated to tricking the world into believing that global warming is a crisis that requires a drastic response."[30]

Because of the nature of these emails, journalist James Delingpole of the UK *Daily Telegraph*, coined the term "Climategate" to describe the scandal, and the name stuck. Other scientists were shocked and dismayed at what these emails revealed. Christopher Booker has called this "the greatest scientific scandal of our age."[31]

After these emails were leaked, other climate scientists such as Dr. Eduardo Zorita and George Monbiot called for Dr. Jones to step down as head of the CRU . But he refused to do so. Instead he blamed skeptics for his refusal to release data while at the same disclosing that he had considered suicide.[32]

James Delingpole, described this collection of emails as "conspiracy, collusion in exaggerating warming data, possible illegal destruction of embarrassing of information, organized resistance to disclosure, manipulation of data, private admissions of flaws in their public claims, and much more."

Other evidence of corruption within the climate alarmist circle has come to light. Testifying before a congressional committee on Dec. 6, 2006, Dr. David Deming, a University of Oklahoma professor, revealed, "I received an astonishing email from a major researcher in the area of climate change. He said, 'We have to get rid of the Medieval Warm Period.'"[33]

The Climategate Coverup

At first this caused quite a stir. And really, the emergence of these emails which revealed such dishonesty and arrogance should have marked the end of the climate alarmism fiasco. But unfortunately the mainstream media covered up this affair, which should have been the biggest story of the year. The Obama administration dismissed this as "irrelevant". So most people have no idea that this important information ever surfaced.

In order to deflect the blame from these corrupt scientists, several media outlets blamed the Russians for hacking their emails. "The computer hack", said a senior member of the Intergovernmental Panel on Climate Change, "was not an amateur job but a highly sophisticated, politically motivated operation." Others went further. The guiding hand behind the leaks, the allegation went, was that of the Russian secret services. James Delingpole called this attempt to deflect from the real story a "glorious red herring". The emails were likely released by a goodhearted soul working at that institution who had access and opportunity.

Most dark legacy media outlets failed to report on this story or gave it minimal coverage, either calling it a "nonstory" or a "distraction from the 'real' climate change debate". The emails were dismissed as having been "taken out of context", "cherry picked" or "fabricated". "The emails are being taken out of context, and the quotes are being misrepresented," alleged Gavin Schmidt in 2009. Despite the obvious indications that the scientific method was not being followed by this prestigious group of scientists, there was hardly any follow-up or investigation. Anyone wondering whether these emails were "taken out of context", should take the time to investigate them more fully.

It appears that the attitude of these scientists was that of activists who were promoting their conclusions rather than scientists impartially analyzing the facts.

The Muir Russell Review and the Independent Climate Change Email Review were criticized for bias and failure to address the most serious allegations. Those whose views differed from the official narrative on this – even independent journalists, skeptics and other scientists who expressed concern - were silenced. The people appointed to investigate this scandal either had little knowledge of climatology or how science works, or they were compromised in some way in a deliberate attempt to hide the truth. The coverup seems almost worse than the original crimes.

Derek Lowe of Science.org commented, "I'm no climatologist, but I'm an experienced, working scientist – so, is there a problem here? I'll give you the short answer: Yes. I have to say that there appears to be several, as shown by many troubling features in the documents that have come out. The first one is the apparent attempts to evade the UK's Freedom of Information Act. I don't see how these messages could be interpreted in any other way except as an attempt to break the law, and I don't see how they can be defended."[34]

A senior editor of *The Atlantic* wrote,

> "I had hoped, not very confidently, that the various Climategate inquiries would be severe. This would have been a first step towards restoring confidence in the scientific consensus. But no, the reports make things worse. At best they are mealy-mouthed apologies; at

> worst they are patently incompetent and even willfully wrong. The climate-science establishment, of which these inquiries have chosen to make themselves a part, seems entirely incapable of understanding, let alone repairing, the harm it has done to its own cause."[35]

You might think that with these shady practices and double dealing laid bare, the public would have demanded an end to this scam. But instead the scandal carefully disappeared from view as if it had never happened. Everything was carefully covered up, and most people still have no idea that these shenanigans had been revealed.

In his book, *Tragedy and Hope,* published in 1966, Carroll Quigley wrote, "The global elites' aim is nothing less than to create a world system of financial control in private hands able to dominate the political system of each country and the economy of the world as a whole.[36]

The great climate change lie is moving the world one step closer to that reality.

---

1. WikiQuote.com/wiki/Maurice_Strong
2. The complete name of his famous book is The Origin of the Species by Means of Natural Selection, or the Preservation of Favoured Races in the Struggle for Life.
3. such as Ted Turner and Bill Gates, who gave a Ted talk calling for a drastic reduction in the world's population.
4. Ball, op. cit., p. 95
5. Steven Mosher interview, Population Research Institute, "Crossroads",Epoch Times 6/30/22
6. Dr. Tim Ball, Human Caused Global Warming: The Biggest Deception in History, Tellwell Talent, 2016, p. 85.
7. Alexander King and Bertrand Schneider, The First Global Revolution, 1991, Pantheon Books
8. Mahajo D. Mesarovic and Eduard Pestel, Club of Rome, *Mankind at the Turning Point,* 1975
9. Edwin Berry, PhD, *Climate Miracle,* 2020, p. 60.
10. Berry, op cit, p. 61
11. Alexander King and Bertrand Schneider, The First Global Revolution, 1991, Pantheon Books
12. Ball, op. cit., p.102
13. Ibid.
14. Ibid.
15. 1995 Science Report, SPM, p. 4
16. Ball, op. cit., p.102

17. Tim Ball, *Human-Caused Global Warming: the Biggest Deception in History*, 2016, cover

18. Zubrin, op. cit.

19. DeWeese, op. cit.

20. Ibid.

21. Costella, Op. cit, pp 29-30

22. Costella, op. cit., p 10

23. Steven Mosher and Thomas Fuller, *Crutape Letters*, 2010, p. 63.

24. Christopher Booker, www.telegraph.co,uk/comment/columnists/christopherbooker/6679082/Climate-Change-this-is-the-worst-scientific-scandal-of-our-generation.html, Nov. 28, 2009

25. Email to Warwick Hughes, Feb. 21, 2005

26. https://www.lavoisier.com.au/articles/greenhouse-science/climate-change/climategate-emails.pdf

27. Ibid.

28. Larry Bell,"Climategate II: More Smoking Guns from the Global Warming Establishment," www.forbes.com, Nov. 29, 2011

29. Costella, op, cit., p. 26

30. Myron Ebell, Director of the Competitive Enterprise Institute's Center on Energy and Environment

31. Christopher Booker, "The Real Global Warming Disaster: Is the Obsession with Climate Change turning out to be the most costly scientific blunder in history?" 2009

32. David Brown, "Climate Scientist Phil Jones Contemplated Suicide over Data Claims", thetimes.com, Feb. 8, 2010.

33. Tony Heller, "Erasing the Medieval Warm Period" https://realclimatescience.com/2018/12/erasing-the-medieval-warm-period/,Dec.23, 2018

34. "Derek Lowe, "Climategate and Scientific Conduct", Science.org, Dec. 1, 1999

35. Clive Crook, *The Atlantic*

36. Carroll Quigley, Tragedy and Hope: A History of the World in Our Time, New York: The Macmillion Co., 1966

Chapter 4

The Role of "Greenhouse Gases"

"What happens in the Arctic may be an indicator of what will happen in the rest of the world. How about what happens in the Antarctic then? Since its ice area has been increasing, is this also an indicator of what might be happening in the rest of the world?" —Dr. Ben Herman, past Director of the Institute of Atmospheric Physics

Carbon Dioxide and Water Vapor

On a sunny day the temperature inside your car is much warmer than it is outside. Just as in a greenhouse, sunlight streaming through the glass is trapped inside as heat. Greenhouses and automobiles, having walls and roofs, are closed systems. But the earth is not a closed system. Its atmosphere has no walls or roof as a greenhouse does. All of the sunlight falling on Earth's surface is radiated back into space. If this weren't true, Earth would just keep getting warmer and warmer. So the earth is not a greenhouse, and thus the terms "greenhouse effect" and "greenhouse gas" are misnomers.

A few atmospheric gases, such as water vapor, carbon dioxide and methane, hold onto some of this incoming energy for a short time before releasing it. Without these gases the earth's average temperature, about 15°C, (59° F) would instead be 28 degrees Celsius colder, about -13°C. (close to 0° F), and life as we know it wouldn't exist. Heat — infrared or long wave radiation — isn't really trapped by these gases as we've been told. They hold onto it for a brief time before it is released back into space.

Also, contrary to what the media and our educational institutions tell us, the main "greenhouse gas" is water vapor, **not** carbon dioxide. Making up about 2% of Earth's atmosphere, water vapor, 20,000 parts per million (ppm), is at least seventy times as powerful in atmospheric warming as is carbon dioxide, which makes up only about 0.04% of the atmosphere: just four hundredths of one percent, or 420 parts per million (ppm). In ideal conditions at the equator sunlight falling on one square meter of surface area produces approximately 1000 watts of power. According to Dr. Richard Lindzen, Professor of Physics at the Massachusetts Institute of Technology, the radiative forcing from this amount of CO_2 makes a difference of only about 1 watt/square meter.

The public is being led to believe that there is a much greater percentage of carbon dioxide in the air and, even worse, that carbon dioxide is the principal greenhouse gas and the temperature control knob. Interestingly, policy makers who were asked during congressional hearings what the percentage of carbon dioxide in the atmosphere is had no idea. People who are making energy policy should certainly understand this simple but important fact.

Water vapor, comprising about 95% of all "greenhouse gases" has been entirely forgotten in this discussion. This is a serious problem and, when being purposely left out, a grievous lie. It is because of this type of lie that people – even the brightest of us – have been so completely misled. This fact is even left out of college science textbooks. If people who write science textbooks don't even know, or are intentionally withholding from others, the simple fact that water vapor is the main "greenhouse gas", we are in serious trouble.[1]

Most CO_2 in the atmosphere is emitted (outgassed) from the oceans and the tropics of South America and Africa, where the sunlight is more direct and water is warmer. CO_2 also emanates from the earth's interior via volcanoes. The earth has about 1500 active volcanoes, 119 of which are under water. 75% of underwater volcanism results from magma emerging from the mantle through cracks that extend about 65,000 km (40,000 miles) where tectonic plates are separating. This carbon dioxide has been stored in the earth's mantle since our planet was formed billions of years ago.

The map below reveals the sources of carbon dioxide. The red and bright yellow areas emitting the most CO_2 are in Africa, which has few industries or vehicle traffic, central South America, which is largely unpopulated, the tropical oceans and southeastern China, which is heavily industrialized. There is much less CO_2 being emitted in the United States, western Europe and Russia.

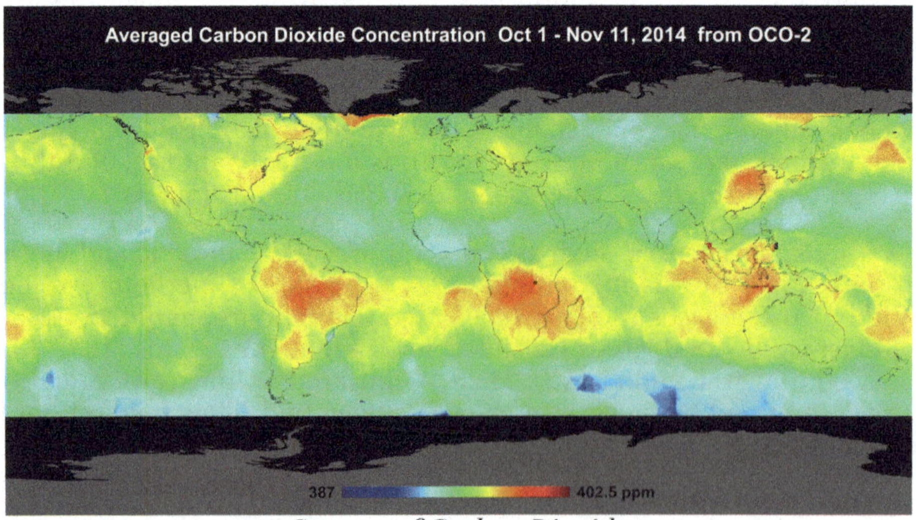

Sources of Carbon Dioxide

Carbon dioxide is also emitted by dead, decaying organisms, termites, bogs and swamps. This exchange between plants, animals, rocks, the atmosphere and the oceans is known as the carbon cycle. A study reported by the New York Times concluded that termites, digesting vegetable matter with the help of bacteria, emit more than twice as much carbon dioxide as all the world's smokestacks.[2]Termite activity has been particularly high due to the widespread clearing of land for raising corn and sugar cane to produce ethanol and for building wind and solar farms. These operations have provided abundant food for termites in the leftover debris.

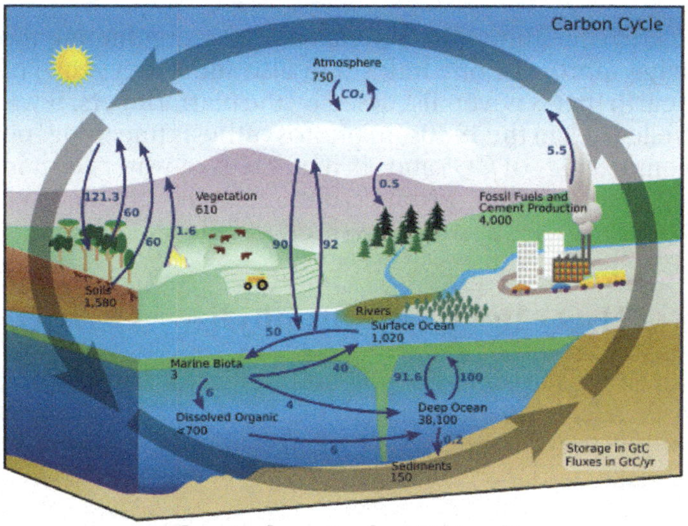

The Carbon Cycle Wikipedia

After carbon dioxide is emitted, it is mixed with other atmospheric gases due to pressure differences and convection currents. The ratio of carbon dioxide to oxygen, nitrogen, and argon is consistent up to about 90 km in altitude. [3]

Photosynthesis, a sink for the absorption of CO_2, is the most important chemical reaction on earth, because it provides all of our food. Plants use it to make food for themselves. Animals then eat the plants, absorbing these nutrients. Carbon dioxide also seeps into the ground and into caves, where it forms calcium carbonate, limestone and clay. Places where carbon is absorbed are called sinks.

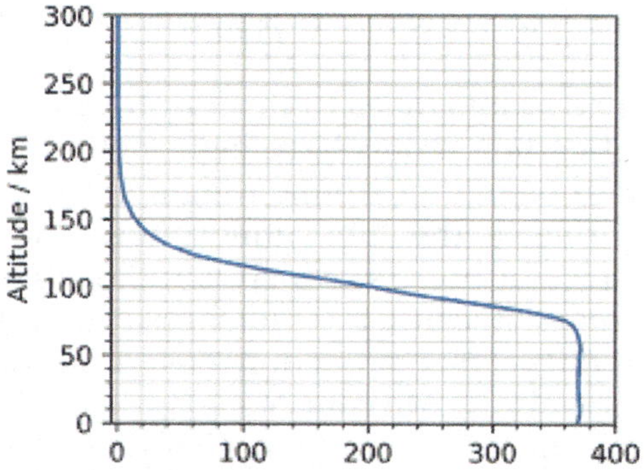

lapper

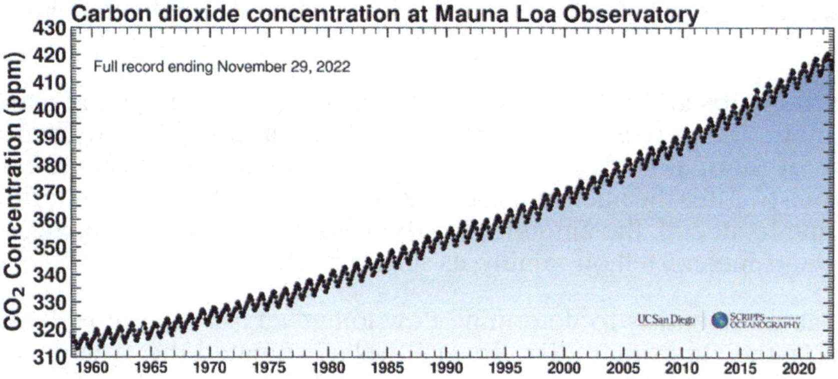

The zigzag line on this graph reflects the fact that the amount of carbon dioxide is lower in the summer when being absorbed by plants in photosynthesis. Each year the amount of CO_2 increases by about 1.5 ppm. The difference from summer to winter is greater than 1.5 ppm, indicating that CO_2 is absorbed rather quickly. This graph also reveals that in the COVID lockdowns of 2021-2022, when people weren't driving or flying much, the amount of CO_2 wasn't any lower. This is further evidence that most of the carbon dioxide being emitted — 95% — is not from humans but from natural sources. Only 5% of atmospheric carbon dioxide is emitted by humans.

With more CO_2 in the atmosphere, plants are now conducting more photosynthesis, producing more carbohydrates and also more oxygen. This is an added benefit to the earth that is never discussed by the fearmongers. It was predicted by Dr. Roger Revelle back in the 1980s. In photosynthesis (meaning "putting together with light") six molecules of CO_2 and six molecules of water react in the presence of sunlight and chlorophyll to produce one molecule of glucose (sugar) and six molecules of oxygen. The reaction is written as follows.

$$6CO_2 + 6H_2O + \text{light \& chlorophyll} \longrightarrow C_6H_{12}O_6 + 6O_2$$

The remaining CO_2 seeps into the soil and bodies of water, joining with other elements such as calcium to form animal shells and rocks. An estimated 39 million gigatons (3.9×10^{16} tons) of carbon is locked up in the shells of ancient animals whose bodies make up the ocean floor.

The Residence Time of Carbon Dioxide in the Atmosphere

Climate alarmist scientists and the IPCC continue to claim that carbon dioxide remains in the atmosphere for hundreds — or even thousands — of years. But this is absolutely incorrect. Several studies have shown that the average carbon dioxide molecule stays in the atmosphere for 5-10 years.[4]

The proof comes from an isotope of carbon: carbon-14. A very small percentage of carbon atoms have eight neutrons instead of the usual six. This isotope of carbon has an atomic weight of 14 (6 + 8) and is called carbon-14. Carbon-14 is formed in the upper atmosphere when nitrogen atoms are bombarded by cosmic rays. Carbon-14, which is radioactive, is present in trace amounts in all living things. Because it is radioactive, it eventually decays back to nitrogen, half of it in 5,730 years. A procedure called carbon dating is used to estimate the age of dead organisms. After an organism dies, it is unable to absorb any more carbon-14. Meanwhile, over a period of years the carbon-14 in the dead organism decays to nitrogen. An organism that has lost half of its carbon-14 has been dead for 5,730 years.

During the 1950s and 1960s, when nuclear testing was being conducted in the atmosphere, the percentage of carbon-14 in the atmosphere nearly doubled, because of thermal neutron reactions in these nuclear explosions, called the "bomb pulse". After the nuclear test ban treaty in 1963, when atmospheric nuclear tests came to an end, the amount of carbon-14 in the atmosphere, measured by mass spectrometers, fell off rapidly, as shown by this graph.

This enabled scientists to determine how long every carbon dioxide molecule remains in the atmosphere. Dr. Roger Revelle estimated the average residence time of carbon dioxide in the atmosphere to be seven years. Despite this proof, the IPCC still insists that CO_2 remains in the atmosphere for hundreds of years.

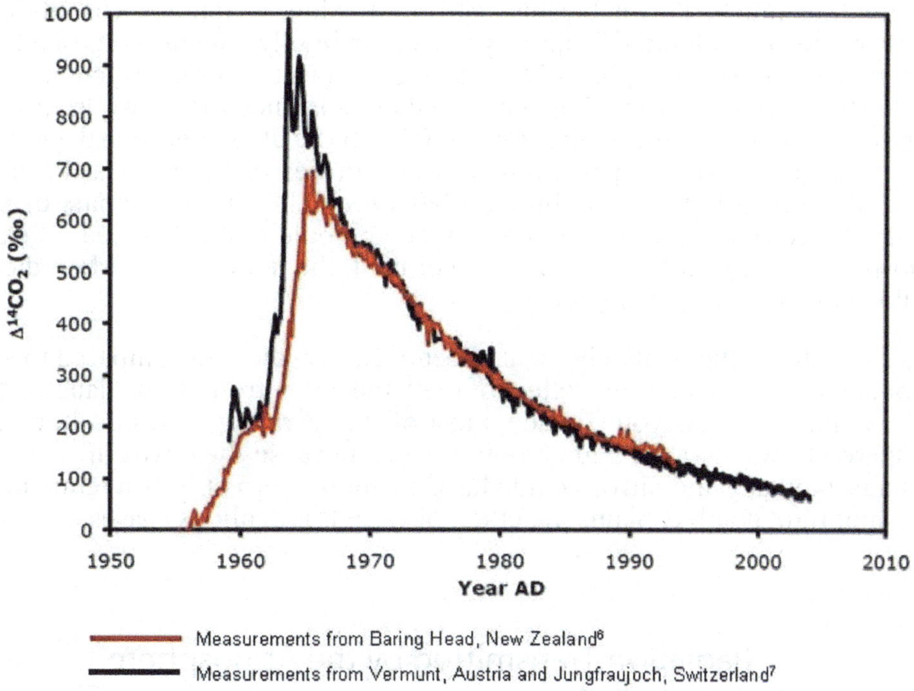

How the Amount of C-14 Fell Off After 1963

The IPCC also erroneously states that CO_2 molecules emitted by the combustion of hydrocarbons are somehow different than CO_2 molecules emitted by volcanoes, termites, and other natural sources. This is ridiculous, because it violates the Equivalence Principle of physics. "The Equivalence Principle states that if data cannot tell the difference between two things, then the two things are identical."[5]There is no way to differentiate between CO_2 molecules emitted by humans and CO_2 molecules emitted by nature. They are exactly identical in every way. Yet the IPCC is telling us that somehow those molecules emitted by humans are different and stick in the atmosphere for hundreds or thousands of years. They peg the increase of CO_2 in the atmosphere from 280 to 420 ppm to have been entirely caused by humans.

The core issue of climate change is: Why has the percentage of carbon dioxide increased? During the 1990s the rate of CO_2 emissions from the combustion of fossil fuels rose nearly linearly by about 0.08 billion tons of carbon dioxide per year. After 2002 the amount of "fossil fuel" CO_2 emissions rose about 50% faster than before. If humans were causing a significant amount of the CO_2 increase, the line in the graph would have been much steeper after 2002. But the Mauna Loa graph shows a linear increase, indicating that the human influence has had virtually no effect on the rate at which carbon dioxide has been rising since 1957. The human contribution to the amount of CO2 in the atmosphere is about 5% of the total. Natural emissions of carbon dioxide are two orders of magnitude larger, over 95%.[6]As suggested in 2008 by Dr. Richard Courtney, most of the 30% increase in CO_2 occurring since the end of the Little Ice Age was caused by its release from the oceans as they have been slowly warming since then. [7]

The Saturation of Carbon Dioxide Molecules

Increasing the amount of CO_2 in the atmosphere from 0 to 280 ppm (parts per million) increases its absorption of heat, called opacity, by about 6.2%. Increasing it from 280 ppm to 420 ppm (where we are now) makes a difference of only 0.6%, one-tenth as much. This is because carbon dioxide, which absorbs and emits energy within a very limited frequency range, can briefly hold photons of infrared light at just two wavelengths: 4.26 micrometers (asymmetric stretching mode) and 14.99 micrometers (bending vibrational mode). Once those two frequencies have been occupied, further additions of CO_2 to the atmosphere make virtually no difference. Dr. Will Happer, physicist and Professor Emeritus at Princeton University, equated this to painting a barn red. The first two coats of paint change the color to red. But the next ten coats would not be noticed. If earth's atmosphere were a 100-meter-long soccer field, the increase in carbon dioxide would amount to just 1 centimeter.

This chart shows the relatively small effect of carbon dioxide compared to water vapor and the truly minuscule effect of methane and nitrous oxide (laughing gas) in the atmosphere. As you can see, most of the warming is from water vapor. Compared to water vapor and carbon dioxide, because they exist in such small amounts, methane and nitrous oxide have an almost zero effect on temperature. This important graph explains the effect of various greenhouse gases.

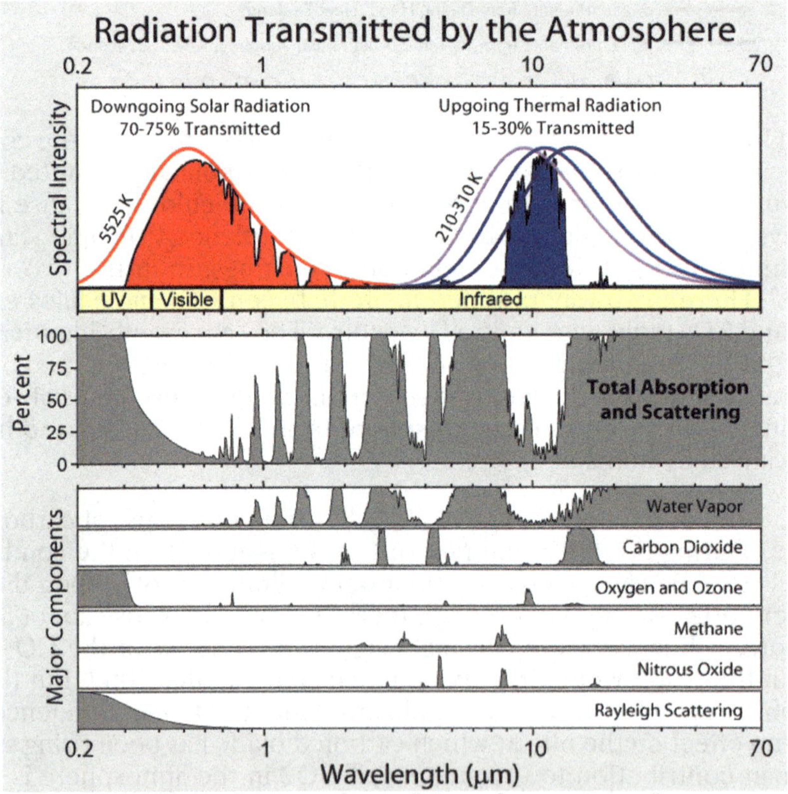

Effect of Atmospheric "Greenhouse Gases"

Should CO_2 rise from its current level of 420 ppm to 690 ppm, the amount of heat it could hold onto would increase by only another 1%.[8] This is because at the current level the wavelengths at which this can occur are saturated. Just as a sponge that is saturated with water can hold no more water, a carbon dioxide molecule can hold no more heat once these two wavelengths in the emissions band are saturated. At preindustrial levels — 288 ppm — carbon dioxide molecules were 81% saturated. At the current level of 420 ppm, they are 87% saturated. The warming effect of each molecule of CO_2 declines as its concentration increases. This fact is acknowledged by both sides of the issue. [9]

The small increase in carbon dioxide we've seen in the past 50 years has been a boon to agricultural production, which has increased by about 15%. As a result the earth has become noticeably greener shown by this composite of satellite photos.

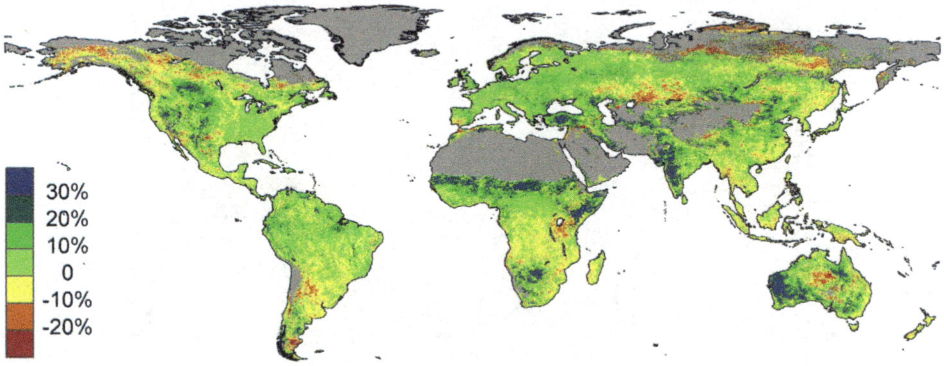

The Greening of the Earth

Plants are growing faster and absorbing more CO_2 in the process. The increased rate of photosynthesis removes more CO_2 from the atmosphere and thus is a negative feedback mechanism. Also, because the atmosphere contains more carbon dioxide, plants don't need to keep their tiny pores known as stomata open as much to take it in. Closing their pores also prevents water evaporating from their leaves. This enables plants to grow in drier places that were previously unable to support them. Satellite photos show the Sahara Desert shrinking.

Yet in December of 2009 the Environmental Protection Agency (EPA) under the Obama administration in the US erroneously classified this life-giving compound as a "pollutant".[10] Since then alarmists have been erroneously using the phrase "carbon pollution" to describe the release of carbon dioxide into the atmosphere. This "Endangerment Finding" was finally rescinded by President Trump on February 12, 2026.

A scientist at the US Department of Agriculture wrote in 1941:

> "Much has been written about varying amounts of carbon dioxide in the atmosphere as a possible cause of glacial periods. The theory received a fatal blow when it was realized that CO_2 is very selective as to the wavelengths of radiant energy it will absorb, filtering out only such waves as even very minute quantities of water vapor dispose of anyway. No probable increase of atmospheric CO_2 could materially affect either the amount of insolation [sunlight] reaching the surface or the amount of terrestrial radiation lost to space."[11]

This graph depicts the levels of CO_2 and temperature for the past six hundred million years. As you can see, there have been times — such as 440 million years ago — when the earth was in an ice age, and the percentage of CO_2 was ten times higher than today. For most of this time the earth's average temperature has been higher than it now is. This graph demonstrates that on a large time scale there is no correlation between temperature and carbon dioxide levels. For the last 300 million years the average amount of carbon dioxide in the atmosphere has been about 1200 ppm, which turns out to be the optimum level for plants. This is the reason why commercial growers pump carbon dioxide into their greenhouses.

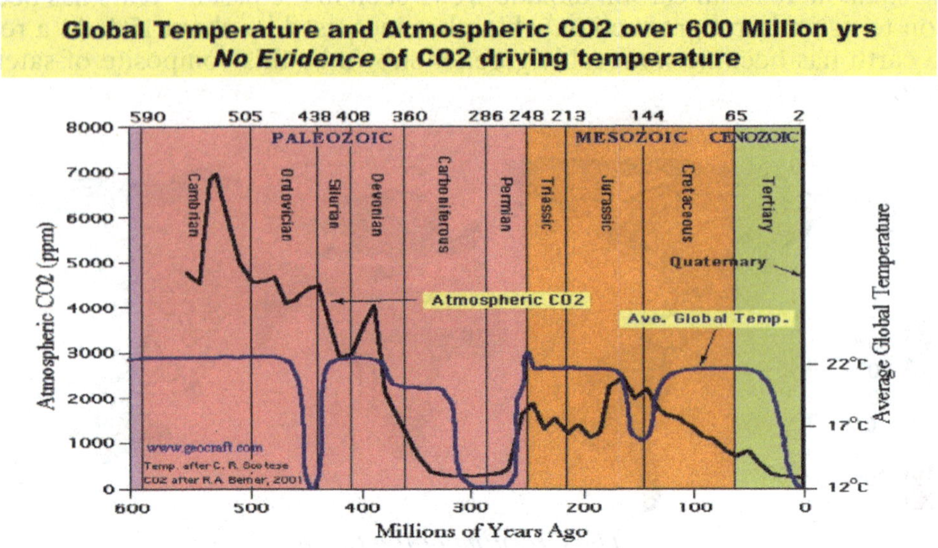

Late Carboniferous to Early Permian time (315 million years ago - 270 million years ago) is the only time period in the last 600 million years when both atmospheric CO2 and temperatures were as low as they are today (Quaternary Period). At no point do temperature and CO2 levels relate. *Temperature after C.R. Scotese http://www.scotese.com/climate.htm CO2 after R.A. Berner, 2001 (GEOCARB III)*

Climate alarmists claim that for the last several thousand years until 1750 the level of carbon dioxide was constant at 280 ppm. This figure was derived from an ice core analysis. But other proxy measurements indicate that the atmospheric level of CO_2 has fluctuated quite a bit, being somewhat higher or lower at various times and locations. Because the level of CO_2 in the atmosphere has risen since then and because Earth's temperature has also slightly increased, they made the assumption that the rise in CO_2 caused the temperature increase. In other words, because there is a correlation between the increases in these two quantities, they believe that the increase in CO_2 caused the temperature increase. This is a logical fallacy, which Aristotle called "affirming the consequent". An example of this would be: "When I get the flu, I have a fever. I have a fever. Therefore, I have the flu."

A new scientific study by Dao Ata et al, published in *The Science of Climate Change*, challenges the assumption that the human emission of carbon dioxide has caused the temperature increase. This study, using multivariate analysis, and multiple linear regression techniques, concluded that the increase in sea surface temperatures (SST) has contributed much more to the temperature increase than carbon dioxide emissions by humans. Multivariate analysis (MVA) is a set of statistical methods of analyzing data with more than two variables. It compared the impacts of sea surface temperatures and human emissions on atmospheric levels of CO_2 .

The sea surface temperature data used in this study was derived from NASA and UK-HADLEY Centre datasets. It found an extremely high correlation coefficient of 0.995 between predicted CO_2 levels found by the UK-HADLEY Centre and actual measurements taken by NOAA. In comparison was a much smaller correlation between human CO_2 emissions and temperature increases. This study also questioned the accuracy of historical CO_2 levels taken from ice core data and multiple measurements taken elsewhere.

Another study, conducted in 2005, determined that changes in the level of CO_2 were not caused by human emissions of this gas and could not explain why its percentage had risen. This agrees with a 2009 study by Dr. Jari Ahlbeck, which declared the correlation between temperature changes and CO_2 to be "clearly statistically insignificant".[12]

Carbon Dioxide is Invisible.

The photos of ominous looking clouds rising from the chimneys of coal burning plants, portrayed in the media, are actually clouds, not carbon dioxide. Whereas you can see clouds — made of tiny drops of liquid water — carbon dioxide is an **invisible** gas which can't be seen in photographs. These misleading photos are sometimes even touched up on Photoshop to increase contrast, making them appear menacing. Water is also a product of combustion.

Some people have also confused life-giving carbon dioxide (CO_2) with carbon monoxide (CO), an extremely toxic gas. We need to get our facts straight.

Water droplets (clouds)

CO_2 and Ocean Acidification

The earth's oceans, covering about 70% of our planet, contain about 40,000 gigatons (billion tons) of CO_2, 50 times as much as the atmosphere holds: 800 Gt (gigatons). When ocean water warms, some of this carbon dioxide comes out of solution. This is because more carbon dioxide can dissolve in cold water than in warm water. The same thing happens when you let a cold glass of soda pop sit until it reaches room temperature. The carbon dioxide that made it fizzy escapes as the liquid warms up. Another glass of pop left in the fridge keeps its fizz (CO_2) for a longer time.

Because oceans have 250 times the mass of the atmosphere, they have a significant effect on climate. Water has a high specific heat. Much more energy is required to change the temperature of water one degree than changing the temperature of soil or a rock. The ocean holds a thousand times more heat than either the land or the atmosphere.

We've been told repeatedly that because of increased CO_2 the oceans are becoming more acidic. Neutral liquids such as water, containing equal amounts of H^+ and OH^- ions, have a pH of 7.0 on the pH (potential hydrogen) scale. Acids have a pH of less than 7.0. Bases have a pH that's higher than 7.0. The pH of ocean water ranges from 7.5 to 8.3. So ocean water is basic, not acidic.

Carbon dioxide reacts with water to form carbonic acid, a weak acid (H_2CO_3). But since the end of the Little Ice Age, about 1750, the ocean has been warming. When CO_2 is released from the ocean into the atmosphere, there's less, not more, carbon dioxide - and thus carbonic acid - in the water. So when ocean water warms, releasing CO_2 into the atmosphere, it becomes less — not more — acidic. This is not to say that a tidal pool into which acidic water from the land is flowing, could have a slightly lower pH. But in general, ocean water could never become even close to acidic.[13]

Ocean Temperatures and Currents

Areas of the ocean that are distinctly warmer than the rest of the ocean are called "warm blobs". James Kamis, a geologist who was interviewed on the Tom Nelson podcast #363 on January 7, 2026, believes that this phenomenon is caused by an excess of geologic activity on the ocean floor in areas such as the "Ring of Fire" surrounding the Pacific Ocean where tectonic plates are colliding and in the Atlantic Ocean, where they are separating.

Many people would be surprised at the number of mountains, including thousands of volcanoes, underneath the ocean. A new high-tech project dubbed GEBCO, which began in 2017, took high resolution images of the ocean floor near the seven-mile-deep Marianas Trench. This project discovered thousands of geysers — all contributing heat from the mantle to the ocean — and many more ocean-floor volcanoes than previously imagined. There is much more geologic activity here than anyone previously suspected.

Yet the IPCC is so hung up on their human-caused carbon dioxide hypothesis that they insist these warm blobs are caused by heat transfer from the atmosphere above the Pacific Ocean and the ocean water. If this were true, the ocean would be warming by the same amount in all areas and there wouldn't be any warm blobs. These warm water areas are also releasing more than an average amount of carbon dioxide, because when water warms, more CO_2 is released (outgassed). Satellites have detected the release of large quantities of helium.

Life-Giving Carbon Dioxide

Dr. Patrick Moore, one of the founders of Greenpeace, an organization that was responsible for stopping nuclear testing in the atmosphere, and the hunting of whales and baby harp seals, came to a very interesting conclusion about the percentage of CO_2 in the atmosphere.

"Human emissions of carbon dioxide have saved life on Earth from inevitable starvation and extinction (at 150 ppm) due to lack of CO_2. To use the analogy of the Atomic Clock, if the Earth were 24 hours old, we were at 38 seconds to midnight when we reversed the trend towards the End Times. If that isn't good news I don't know what is. You don't get to stave off Armageddon every day.

"I issue a challenge to anyone to provide a compelling argument that counters my analysis of the historical record and the prediction of starvation based on the 150-million-year trend. . .I submit that much of society has been collectively misled into believing that global CO_2 and temperature are too high when the opposite is true for both. Does anyone deny that below 150 ppm CO_2 plants will die? Does anyone deny that the Earth has been in a 50-million-year cooling period and that this Pleistocene Ice Age is one of the coldest periods in the history of the planet?"[14]

Dr. Moore points out that the percentage of carbon dioxide in the atmosphere had been declining for the last 150 million years: from 6,000 ppm [parts per million] to just 180 ppm. Since plants need at least 150 ppm of CO_2 in order to survive, they will die if the percentage dips lower than this. Since plants also feed animals, we are talking about not just plants but **all** life on earth dying out if the percentage of carbon dioxide is reduced to a level lower than 150 parts per million.

Methane and Climate Change

Methane (CH_4) has also received an unfair rap from the prophets of doom. One article states: "Due to the chemical bonds within its molecule, methane is much more efficient at absorbing heat than carbon dioxide. . .making it a very potent greenhouse gas. Fortunately, methane can only last about 10 to 12 years in the atmosphere before it is oxidized and turns into water and carbon dioxide."[15]

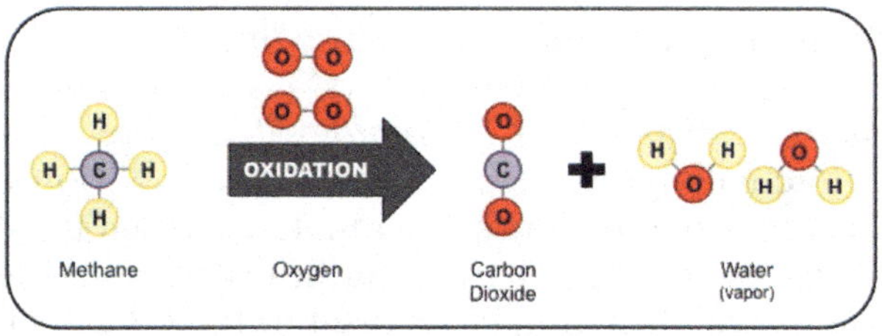

Because each molecule of methane, or natural gas, has more efficient heat trapping qualities than CO_2, it is now listed as a "dangerous" gas. In 1995 the IPCC declared it to be 21 times as strong as carbon dioxide. What they haven't told you is that methane makes up just 0.00002% (two parts per million) of the atmosphere — totally insignificant. This is just another huge scare tactic. Nitrous oxide (N_2O) is found at just 0.3 parts per million. Please take a look at the effect of methane and nitrous oxide on the chart shown earlier in this chapter. Obviously the propagandists are assuming that the average person doesn't understand numbers and will never question their frightening pronouncements.

US President Joe Biden called climate change "an existential threat" and "the number one issue facing humanity". Yet on February 7, 2022, he warned, "If Russia invades [Ukraine], then there will no longer be a Nord Stream 2. We will bring an end to it." The attack on the Nord Stream Pipeline in 2022 released 445,000-485,000 **tons** of methane gas, making it perhaps the largest human-caused release of methane in history.

Climate, Farms and Meat Eating

You might have heard that farm animals are a significant source of methane, a "greenhouse gas". The fact that ruminants, such as cows and sheep, produce methane as a product of their digestive processes is cited as a reason to reduce meat consumption. But it's important to look at the big picture. Compared to other sectors, such as industry and transportation, the farm's contribution to the warming of the atmosphere is miniscule. Many farmers practice regenerative, sustainable farming methods that sequester carbon and improve soil health. These practices offset methane emissions from livestock.

What about all the methane being produced by wildlife? Should we also get rid of lions, tigers, elephants, rhinos and deer and other wild animals because they, too, are producing methane?

Under the leadership of the World Economic Forum (WEF), governments have been putting farmers all over the world under intense pressure to reduce the amount of methane and nitrous oxide their ruminant animals are producing. Using this excuse, the government of the Netherlands has stolen thousands of farms to "save the planet". Some of these farms have been in the same family for generations. These unjust actions are adding to impending food shortages, which could cause millions of people to starve.

Because methane is also produced in the growing of rice, some policymakers want to forbid farmers from growing this crop that is a staple of half of the world's people.

Many environmental activists are pushing for a shift away from meat consumption and towards lab-grown meat, also known as cultured or cell-based meat. But is this the solution it's made out to be? Lab-grown meat is produced from immortalized cell lines, similar to cancer cells that are also "immortal". These are multiplied in the lab using nutrients and growth factors. A recent study published in the journal *Environmental Science and Technology* discovered that the production of lab-grown meat actually generates 2.5 times more carbon dioxide per gram of protein than conventional meat production. Lab-grown meat production also requires a lot of energy, primarily from natural gas, which emits CO_2 when burned. The nutrients and growth factors used to feed these cultured cells have their own emissions footprint, contributing to the overall CO_2 impact.

In 2017 Bill Gates invested $17 million in Memphis Meats, a lab-grown meat startup. The Bill & Melinda Gates Foundation has provided funding to various organizations working on alternative protein sources, including plant-based and cultured meat.

Obviously this policy was motivated by aggressive, militaristic vegans who won't be satisfied until they have captured the entire world with their lifestyle. They'll insert any excuse they can think of to accomplish this, even making up lies. They've made the following pronouncement: "Emissions from cows are more damaging to the planet than all the planes, trains and automobiles combined." But strangely, at the COP-30 Climate Change Conference held in Belem, Brazil, in November of 2025, meat was served at several restaurants at the conference center, and it was reportedly gone in two days. This conference, which accomplished very little if anything, has received a lot of criticism. An eight-mile stretch of highway pushing through the Amazon rain forest was built just for this conference. Many of its thousands of participants, many of whom had arrived in private jets, were housed in giant cruise ships docked offshore.

Arctic peat bogs that are contained in permafrost would release methane if the permafrost melted. However, a study on peat mosses in the Netherlands revealed that a large percentage of this methane is eaten by methanotrophs, bacteria that consume methane.[16]One amazing thing about methanotrophs is that they produce PHB, a biodegradable plastic that is not toxic. Products made with PHB can be placed in landfills. This amazing technology is being studied at the University of Alberta in Canada.

The US and the European Union pressured over 100 nations to reduce the amount of methane and nitrous oxide in their countries by a whopping 30% by 2030. How will they do that unless they kill the farm animals and decrease the amount of rice grown? If these nations actually go through with this travesty, millions of people will starve. This has already happened in Sri Lanka, Ghana and South Africa.Despite the leaks of methane and the fact that farm animals breathe it out, its existence is still only **2 ppm** (compared to 420 ppm of CO_2). Vilifying these gases is just another devastating trick perpetrated by the climate change industrial complex.

1. Konrad B. Krauskopf and Arthur Beiser, *The Physical Universe, 13th Edition*, McGraw Hill, 2010, pp. 111-112.
2. https://www.nytimes.com/1982/10/31/us/termite-gas-exceeds-smokestack-pollution.html
3. Dr. William Tapper, and % of CO2 chart
4. Dr. Fred Goldberg lecture, (YouTube video) March 15, 2016
5. Dr. Ed Berry, *Climate Miracle*, 2020, p. 19.
6. Professor Murry Salby lecture, July 18, 2016, University College London.
7. Ed Berry, *Climate Miracle*, 2020, p.40.
8. https://stevengoddard.wordpress.com/2016/04/17/global-warming-scam-has-been-dead-for-75-years/
9. The IPCC has acknowledged this, but not in "Summaries for Policymakers".
10. http://dailycaller.com/2015/02/11/epa-challenged-carbon-dioxide/
11. Prof. Murry Salby, "Climate Scientist Murry Salby Demolishes the Global Warming Alarm," lecture on YouTube, Aug. 13, 2014
12. Ibid.
13. Dr. Don Easterbrook, lecture before Washington state senate committee, March 26, 2013
14. https://www.technocracy.news/former-president-of-greenpeace-scientifically-rips-climate-change-to-shreds/
15. Frederick Beaudry, "Methane, a Powerful Greenhouse Gas," Treehugger,11/25/19
16. Nardy Kip, et al, https://www.nature. com/articles/ngeo939/

Chapter 5

Empirical Science vs Climate Science

> "If it disagrees with experiment, it is wrong. In that simple statement is the key to science. It doesn't make any difference how beautiful your guess is. It doesn't make any difference how smart you are, who made the guess, or what his name is. If it disagrees with the experiment, it's wrong. That's all there is to it." —Professor Richard Feynman, Nobel Laureate

Science is more than just knowledge about our natural environment. It's also a way of thinking and approaching problems. If the rules for scientific investigation are not followed, the science is flawed, and the conclusions reached are most likely erroneous. Science isn't omnipotent; it has its limitations. And scientists aren't little gods. They are human beings with the same tendency to be corrupted as everyone else. A true scientist is humble enough to admit s/he was wrong about something. The hubris we are seeing today among global warming alarmists is antithetical to science. "It's almost like a religious creed. Science isn't faith. It's rigorous skepticism." [1]In the COVID-19 dilemma — a similar situation — United States senator Ron Johnson correctly observed, "Research grants have corrupted medical science." This is also true of climate science. And those in power have been censoring anyone who disagrees with the official position.

Is there any way to check whether human-caused climate change has ever been scientifically verified? Just as important: is there any way to **disprove** this hypothesis? The following have been blamed on climate change: more heat waves but also more cold spells, more droughts but also more floods, less snow but also more snow, more mental illness and suicides, more alcoholism, more crime, more bugs, rats and bats, the civil war in Syria, more earthquakes and tsunamis.

A scientific hypothesis is an idea that guesses the cause of some observed phenomenon of nature. Karl Popper (1902–1994) once stated, "If a theory can be falsified, it is science. Otherwise it is pseudoscience [false science]." Since whatever happens with the weather is now blamed on "climate change," there is no way to disprove this hypothesis. If there is a heat wave, we are told that it is caused by climate change. If there is a severe cold snap, that is also blamed on human-generated climate change. Thus it is impossible to show that human-caused (anthropogenic) climate change is not happening, rendering the anthropogenic global warming (AGW) hypothesis invalid.

The burden of proof that human emission of CO_2 is causing climate change rests with those who have been calling this hypothesis a fact for the past forty years, since they made this claim in the first place. To this date absolutely no evidence has been provided that the trace gas carbon dioxide is a significant driver of climate. Have you noticed that this issue is never debated? In 1998 the Global Warming Petition project offered a $10,000 reward to anyone who could provide this evidence. This reward was intended to encourage scientific inquiry into the causes of climate change and to challenge the prevailing narrative. Despite the reward offer **28 years ago**, no one has yet successfully obtained the prize. The lack of applicants suggests that evidence supporting the AGW hypothesis is inconclusive or entirely missing.

Piers Corbin, a British meteorologist noted for the accuracy of his long-range weather forecasts, stated, "There are no scientists in the world who can produce such observational data from the real world. . .There is only effect the other way, namely that the ocean temperatures control average CO_2 levels due to basic laws of physics about the solubility of CO_2 gas in [sea] water." [2]

Empirical (evidence-based) science records data (information) obtained from observations and scientific instruments. Modern thermometers were not in use until the 1800s. In order to learn about Earth's climate in the years previous to this, we have to rely on historical records and indirect measurement, called proxy data. Proxy data are taken from sources such as ice cores, tree rings, glaciers, the tree line and sediment cores from the bottoms of lakes and oceans. These help scientists understand and reconstruct climate patterns going back millions of years. They can be compared (calibrated) with fossils and with data taken within the last hundred years to determine their accuracy. Even though they will never be as accurate as direct observations, they are very useful.

A graph of climate history plotting the amount of CO_2 with the average temperature shows that a rise or fall in temperature always occurs **before** the rise or fall of carbon dioxide levels. Proxy data (other than direct measurement by thermometers) taken from ice cores drilled in Antarctic and Greenland glaciers reveal the temperatures and percentages of various gases at different times in geologic history to around 900,000 years ago. Vostok is a frozen lake in Antarctica.

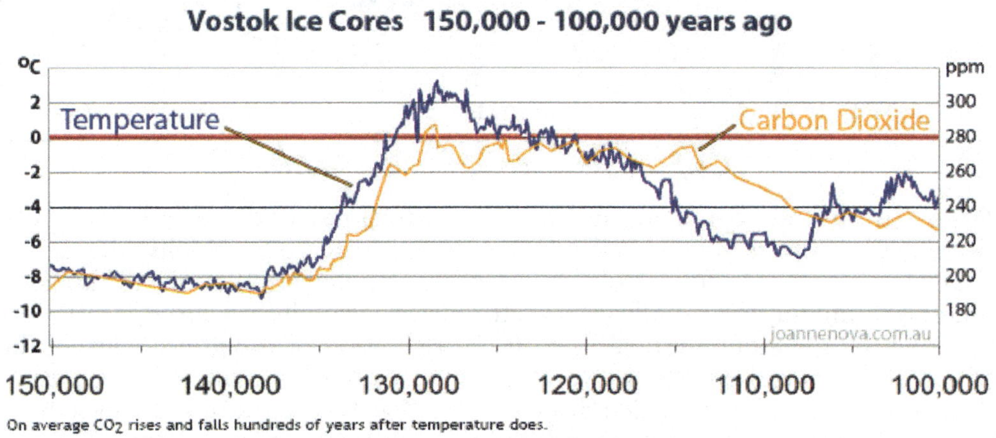

Temperature and CO2 Levels

Data-gathering technology in ice core samples obtained before 1999 lacked the precision to determine this relationship between CO_2 and temperature. Improvements in this technology clearly show that increases in temperature **always** precede increases in CO_2. This fact, known by all climatologists by the year 2003, demonstrates that an increase in CO_2 is **the result, not the cause** of temperature increases.

The average lag time between a temperature increase and an increase in CO_2 is around 800 years, as demonstrated by the above graph. Seventy percent of the earth is covered by oceans. The 800-year lag time is caused by the fact that water, because of its high specific heat, cools and warms up much more slowly than land does. So it takes about 800 years for the ocean to catch up to naturally-caused temperature changes and warm, releasing CO_2. Natural sources of carbon dioxide release about 780 billion tons of CO_2 per year. In contrast, human emissions of CO_2 total 36 billion tons, only about 5% of the total amount of carbon dioxide being released.

The graph showing that temperature changes precede changes in the percentage of carbon dioxide debunks the hypothesis that increases in carbon dioxide produce increases in temperature. That is the reason why careful measurements are absolutely critical to any scientific investigation.

The graph below shows that in the last 420,000 years there have been four periods of glaciation — one about every hundred thousand years, when glaciers covered much of North America and Europe. This last ice age, the Pleistocine, has existed for the last two million years. The periods of time between periods of glaciation are called interglacials. We currently find ourselves in one such interglacial, the fifth in the past 420,000 years.

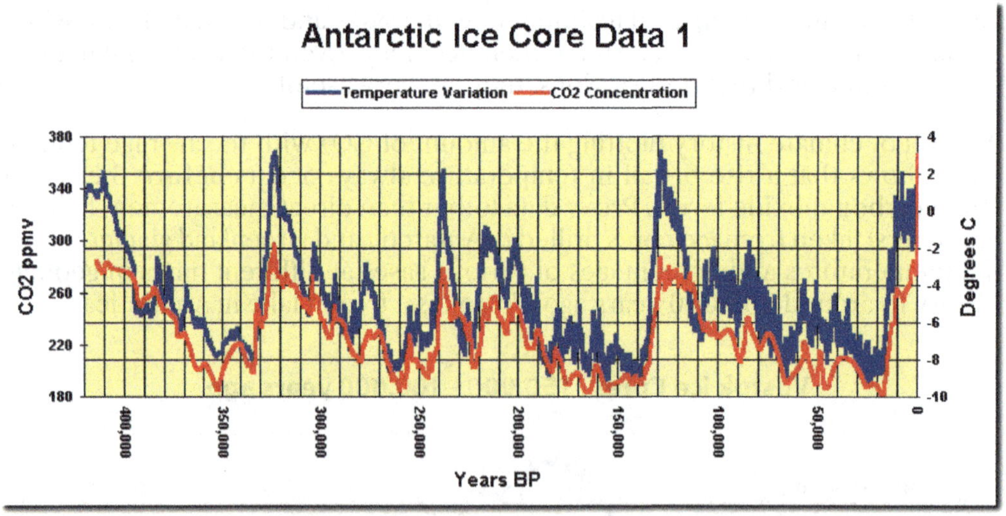

Temperatures precede changes in CO2

Twenty thousand years ago, during the southernmost advance of the Arctic glaciation, northern Europe and North America were covered with almost two kilometers of ice. At the end of the last period of glaciation about 18 thousand years ago temperatures rapidly increased, and the ice melted in about 7,000 years, as shown by the steep vertical line on the right side of the graph. During this period of warming there was a severe cold snap, called the Younger Dryas period, about 12,500 years ago, in which the earth's average temperature suddenly cooled by several degrees for an unknown reason. Then the warming resumed.

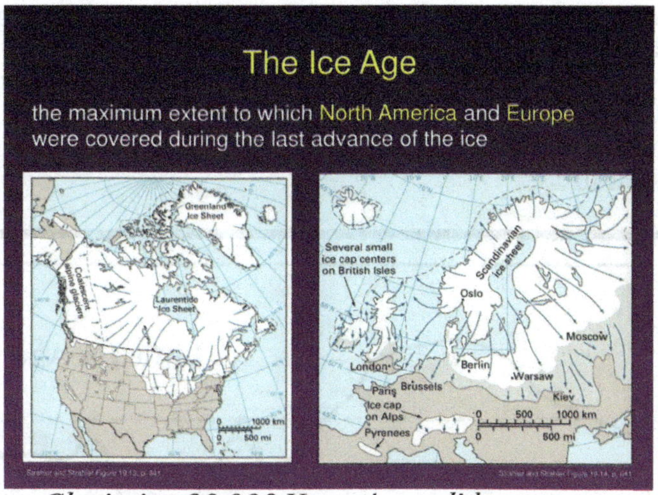

Glaciation 20,000 Years Ago: slideserve.com

Ice core data from Greenland's glaciers also indicate that during most of the past 10,000 years Earth's temperatures were warmer than today. Eight thousand years ago, during what's known as the Holocene Maximum, the Arctic was ice free. The present time is on the left side of the graph below.

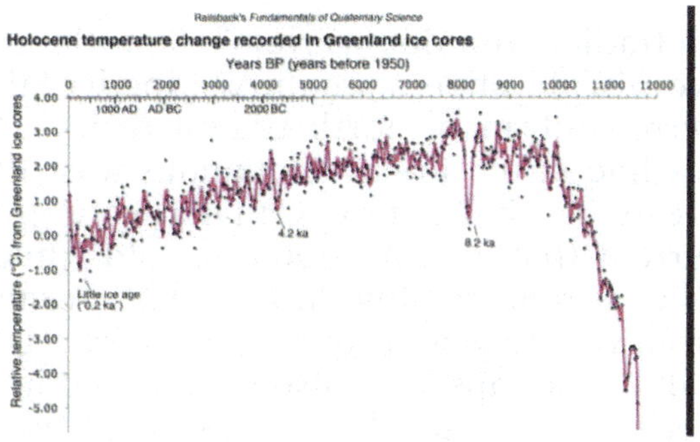

The graph below, constructed from ice core data, shows the Bronze Age Warming, the Roman/Iron Age Warming, the Medieval Warm Period, the Little Ice Age and the Modern Warming.

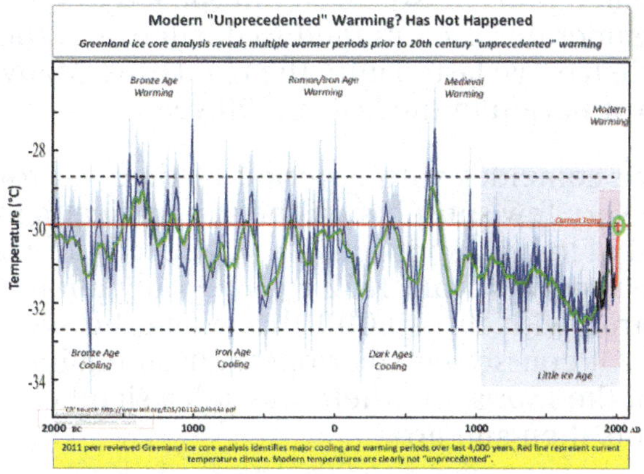

Temperature Record from 2000 BC

The period that occurred after the Roman Warming, from 400 AD to 950 AD, known as the Dark Ages, was colder. Then came the Medieval Warm Period, extending from about 950 AD to 1250 AD. Ice core samples show that although the average temperature was about two degrees warmer during the Medieval Warm Period, the percentage of carbon dioxide was lower than it is now. At that time a colony of Vikings from Scandinavia raised sheep and vegetables in a part of Greenland now covered by a glacier. This again demonstrates the lack of correlation between the percentage of carbon dioxide and Earth's average temperature.

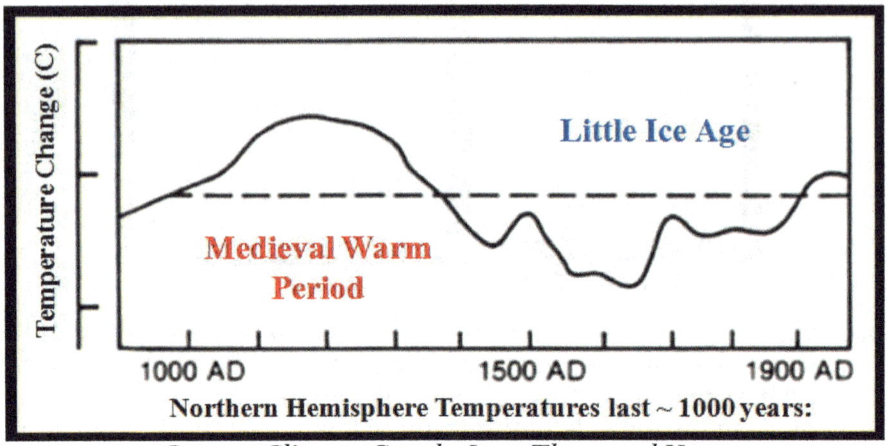

Correct Climate Graph: Last Thousand Years

Empirical data contradicts the doomsayers' claim that the Medieval Warm Period wasn't a worldwide phenomenon. Wikipedia falsely states that this phenomenon occurred only in the north Atlantic region. However, a thousand years ago the snow line in the Rocky Mountains was 1000 feet higher than it is now, and forests in what is now Canada were growing north of the location of the current tree line. A 5000-year-old white spruce was found 100 km north of the current tree line in Canada.[3] Permafrost, in which the soil never melts from year to year, prevents trees from growing north of that line. 1000-year-old tree stumps have also been found in ice caves under the Mendenhall glacier near Juneau, Alaska.[4] A forest was growing there 1000 years ago when it was warmer. This proxy data proves beyond any doubt that temperatures a thousand years ago were warmer than today's.

Then, after about 1300 AD temperatures began to fall again, and by 1350 AD the Vikings' Greenland settlement had been abandoned. Shortened growing seasons and torrential rains in Northern Europe led to the "Great Famine" of 1315-1317 AD.[5] Temperatures continued to decline, and the very cold period known as the Little Ice Age lasted until the late 1700s. Many scientists believe this was the coldest period in the last 10,000 years.

The Little Ice Age coincides with a period of time — from 1645 to 1715 — when there were very few sunspots on the sun, indicating very low solar activity. This period of low sunspot activity was called the Maunder Minimum. During the 28-year period from 1672 to 1699 only about 50 sunspots were observed, compared with the 40,000 that would be seen in a comparable time period now. Of course our solar observation equipment today is vastly superior to that of the 1500s. But there was still a significant difference in the number of sunspots then and now.

Ever since the coldest part of the Little Ice Age, about 1680, the earth has been in a gradual warming trend. Average temperatures have increased about 0.5°C. per century ever since, and scientists really can't explain why. Glaciers began melting around the year 1800.

In the past 500 years there have been about twenty periods of warming and cooling, four of which occurred during the last century.

1. Between 1890 and 1940 average temperatures increased and were actually warmer than they are now. Temperatures in the Arctic were higher during the 1930s than at present, and glaciers were rapidly melting. At that time, however, the level of CO_2 in the atmosphere was only 0.032%, 320 parts per million (ppm), compared to about 420 ppm today.

2. Between 1940 and 1975 Earth's temperatures were decreasing at the same time that CO_2 levels were rising. Arctic ice rebounded during this period of cooling. How could this have happened if carbon dioxide was really the climate control knob?

3. Then, from 1978 to 1998 another period of warming occurred. CO_2 was also increasing, causing some scientists to mistakenly conclude that this rise in carbon dioxide was causing that temperature increase. But they evidently failed to note that the sun was unusually active during that 20-year period of time.

4. Between 1998 and 2023 average temperatures remained fairly constant, even though about 30% of all human generated CO_2 was emitted during this time period.

Warming and cooling during these four time periods counter the AGW hypothesis. The decade of the 1930s, the warmest in the last century, occurred when the carbon dioxide level was much lower than it now is. In the 1960s and 70s the earth was cooling while the CO_2 level was rising. 2013 was the coldest winter in 100 years in Europe. Five of the six snowiest winters in United States history occurred within the last 15 years.

For the first fifteen years of this century the IPCC, CRU and other agencies acknowledged that there had been no increase in temperatures since 1998. Nevertheless, to everyone's surprise, in 2015 a paper by Tom Karl of NOAA, apparently timed to coincide with the upcoming Paris conference, claimed that the temperature data showing this lack of warming never existed and that temperatures had been rising during that time.

Dr. John Bates, a respected NOAA climate scientist, came forward to reveal that this paper, which became known as the "Pausebuster Paper", was based on unverified and misleading data. Nevertheless, Karl's paper had a tremendous influence on policy makers from several nations who were trying to decide how to commit their governments to dealing with what they were told was a crisis.[6] These people trusted the "experts", not realizing that other experts disagreed with them.

The period from October, 2018, to April, 2019, was the coldest and wettest 6-month period in the United States in over a century.[7]

Both RSS (remote sensing system) satellite and ARGO ocean temperature measurement reveal that in the first 18 years of the 21st century temperatures remained fairly constant and may have even cooled a bit. By 1999 evidence surfaced that the amplification of the effects of CO_2 by water vapor originally predicted by climate alarmists was not happening. This fact was finally released in an obscure science journal in 2006 but was ignored by the climate alarmists.[8]

You can now investigate the climate record for US reporting stations using a graphing program, RealClimateTools.com., by Tony Heller. Here you can access data from thousands of weather stations going back to the year 1900. You can look at the trend line for any of these stations to determine whether the temperatures in that community have been rising or decreasing. You can also observe the paths of tropical storms and hurricanes as well as their wind speeds for various years.

A team headed by Dr. Richard Lindzen of the Massachusetts Institute of Technology (MIT) conducted a 2009 study showing that when the earth receives more heat from the sun, it simply radiates the additional heat back into space.[9]

Another climatologist, Roy Spencer, reported in 2007 that during periods of warming there are fewer high clouds to hold in heat.[10] This causes cooling, and acts as a buffer that counteracts feedback from increased water vapor that climatologists were predicting in the 1980s.

Clouds are really the key to understanding climate. They can either hold in heat or block sunlight from reaching Earth's surface, cooling it. This depends on their height. High clouds generally slow down the dissipation of heat into space, causing warming.

On the other hand, low clouds block sunlight, which is then reflected back into space. This reflectivity is called the albedo effect. Albedo is the reflection of sunlight by ice, snow and clouds, which are light-colored surfaces. If more low clouds than high clouds form as a result of extra humidity caused by CO_2, the earth's temperature is cooled. This is a **negative** feedback which offsets the slight warming effect of CO_2.

A 2024 study by Ned Nikolov and Karl Zeller concluded: "The sun's total solar irradiance and albedo fully explain all observed global warming since 1999. Albedo changes are by far the dominant driver of the earth's global temperature. There is no evidence that human carbon dioxide emissions or changes in the carbon dioxide level affect the earth's global temperature."

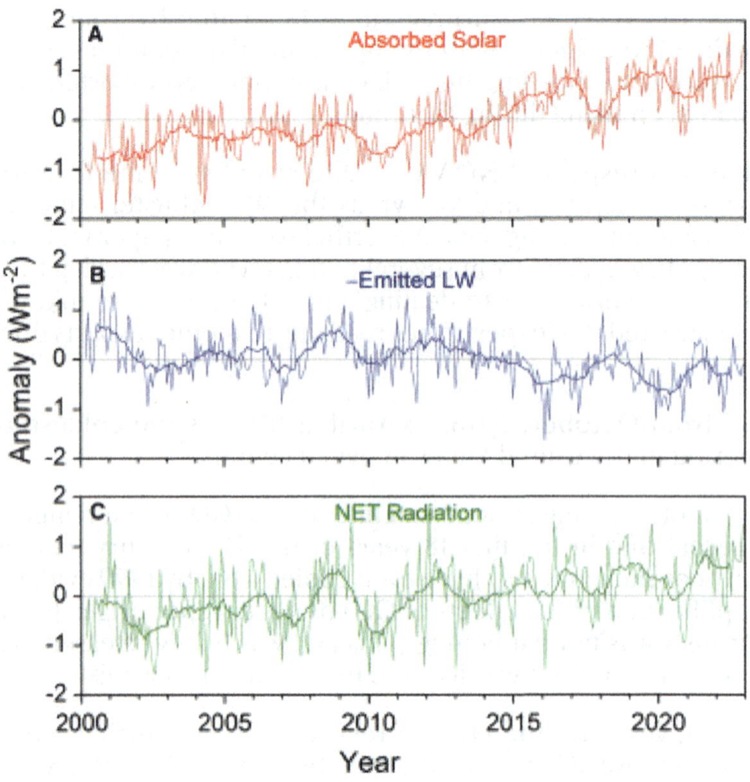

The above graphs depict the influence of clouds in affecting the amount of solar radiation reaching the earth's surface. In the past two decades the amount of cloud cover has decreased slightly, resulting in more sunlight absorbed (A). Thus, slightly less heat, longwave radiation (LW) is reflected (B). The net radiation is the combination of these two factors.

In the 1980s computer models predicted that if CO_2 was causing warming, there would be a "hot spot" over the tropics, called a global warming "signature". The idea was that the resulting warm water vapor in the tropics would rise to a higher level of the troposphere, the lowest level of the atmosphere, about six miles up. For decades thousands of radiosonde balloons sent aloft in that area have failed to detect this hot spot. Many climatologists feel certain that if that hot spot existed, it would have been detected by now.[11]

There are currently five different temperature datasets: two sets of satellite temperature records from the University of Alabama in Huntsville (UAH) and three land-based datasets: Hadcrut in England, NCDC and the NASA GISS temperature dataset. Anyone can download this data from the Internet, put it onto an Excel spreadsheet and run a trend line through the data.

Satellite data, used since 1979, have proved to be a more accurate indicator of the earth's average temperature than surface weather stations for several reasons. First, thermometers used in ground-based weather stations are often located in cities close to air conditioner vents that spew warm air as well as in airports and other paved areas, where concrete, absorbing radiation from the sun, raises the temperature. A century ago, most weather stations were located in grassy, unpaved, rural areas.[12] The influence of human activities such as traffic, concrete and air conditioner exhaust vents in cities is known as the Urban Heat Island (UHI) effect. This can make a difference of up to several degrees in temperature measurement, skewing the results. Cities grew up around weather stations that had originally been in rural areas, but these thermometers were kept in their original locations.

A team of volunteers headed by meteorologist Anthony Watts checked US weather stations to spot those with such problems. When he confronted NOAA about problems at these weather stations, they refused to fix them. They countered that they were compensating for these problems by making "adjustments". But how would they know how much of an adjustment to make at each station? This is sloppy measurement and certainly not science.

About three percent of the earth's land area is now affected by the Urban Heat Island phenomenon. Some warming in selected areas has been caused by changes in land use, especially the destruction of the rain forests and the growth of cities. The average temperatures of the earth's great cities, such as New York, have risen significantly in the past century. Global warming is indeed happening in large cities. Yet the average temperatures of many smaller cities and towns have remained flat or even lower during the same time period.

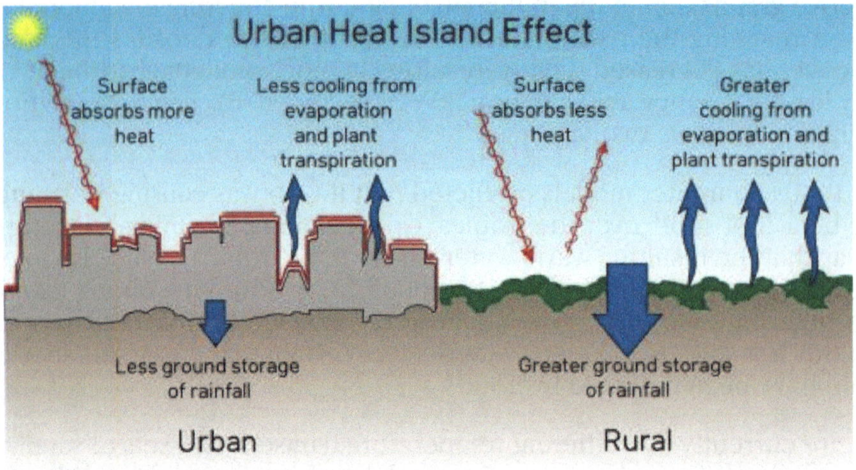

Temperature records have been kept for only about 140 years. Proxy methods of estimating temperature, such as tree rings and ice core samples, reveal the many temperature changes occurring during Earth's climate history.[13] In contrast to land based thermometers, satellite temperature sensing equipment measures all of the earth's surface equally. The ARGO system to measure ocean temperatures wasn't used till 2004.

Another reason that surface weather stations can't measure Earth's average temperature accurately is that they're not evenly distributed around the globe. Most of them are located in the eastern United States and western Europe. There are far fewer weather stations in the southern hemisphere, the Arctic region and on oceans. So in computing Earth's average temperature, many temperatures in those regions are simply guessed at. Land based weather stations cover less than 25% of the earth. Thus, statements that we just had the "warmest year on record", when the average temperature has been recorded as *a* few hundredths of a degree higher than the previous estimate are meaningless. They simply indicate that temperatures were flat.

Even though there were only about 1,000 weather stations used to provide data for the computer models, this number was reduced in the 1960s. Then in the 1990s that number was further decreased, The main concern here is which stations were retired. In Canada most of the Arctic stations were taken out with the only one remaining: in Eureka. Each station is supposed to represent a radius of 1,200 km (745 miles).

This problem is just as serious in many other countries. Consider the Sahel region in Africa. "One obvious problem is a lack of data. Africa's network of 1,152 weather watch stations, which provide real time data and supply international climate archives, is just one-eighth the minimum density recommended by the World Meteorological Organization (WMO). Some of the stations that do often fail to report."

Arctic Weather Stations in Canada

Another serious problem with ground temperature recording station data is the fact that some of these numbers have been made up. Ray Sanders, a retired construction engineer in the UK, discovered through an FOI request that the UK Met Office (Meteorological Office) has been fabricating data from over 100 nonexistent weather stations. He visited the places where these stations were supposedly located and couldn't find them. A large proportion of these had been closed for fifty years. When he confronted the Met Office about this, their answer was that the stations being used were "not an output of the process". Approximately one out of ten stations were in kitchen gardens surrounded by walls; they would naturally be giving elevated numbers due to their locations near buildings. [14]

The United Nations commissioned the IPCC to study average temperature increases due **only** to the carbon dioxide emitted by human activities. They made no allowance for natural causes of climate change: cyclical changes in solar activity, changes in the earth's orbit caused by Jupiter's gravity, or in the tilt of Earth's axis of rotation and the precession of the equinoxes. Just as a spinning top wobbles, the earth's tilt at a particular time in geologic history can range from 21 to 24 degrees over a period of about 41,000 years. The tilt of the earth's axis is what causes seasons. The greater the amount of tilt the greater difference there is between winter and summer.

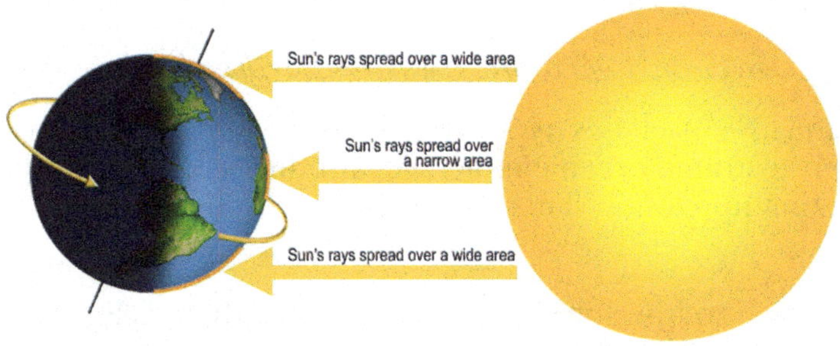

Earth's Tilt Causes Seasons

The earth's orbit around the sun is a slight ellipse. At this time in geologic history we are closer to the sun in early January when the northern hemisphere is facing away from the sun. Because of this, winters in the northern hemisphere aren't as severe as they would be if the earth was farther from the sun in January.

But at other times in a 100,000 year cycle the opposite case occurs. During those periods the earth is farther from the sun when the northern hemisphere is having winter. Northern hemisphere winters are then colder. During those periods Arctic ice wouldn't have as much opportunity to melt and could start accumulating from year to year, leading to a period of glaciation.

Clouds and Cosmic Rays

Since the 1990s Dr. Henrik Svensmark and his team have been studying the connection between the solar wind and cosmic rays. Sunspots, huge magnetic storms on the sun's surface, cause an increase in its magnetic field during the peaks in its 11-year sunspot cycles. When there are more of these magnetic storms on the sun, it releases more charged particles, called the "solar wind", into space. During those times more of these particles reach Earth. The solar wind, along with Earth's magnetic field, deflects cosmic rays heading toward Earth from outer space. So when the sun is more active, fewer cosmic rays are able to reach Earth's surface

Cosmic rays are not electromagnetic rays but highly energized atomic-sized particles entering Earth's atmosphere at nearly the speed of light. These are believed to originate from huge explosions called supernovae, in which massive stars end their lives.

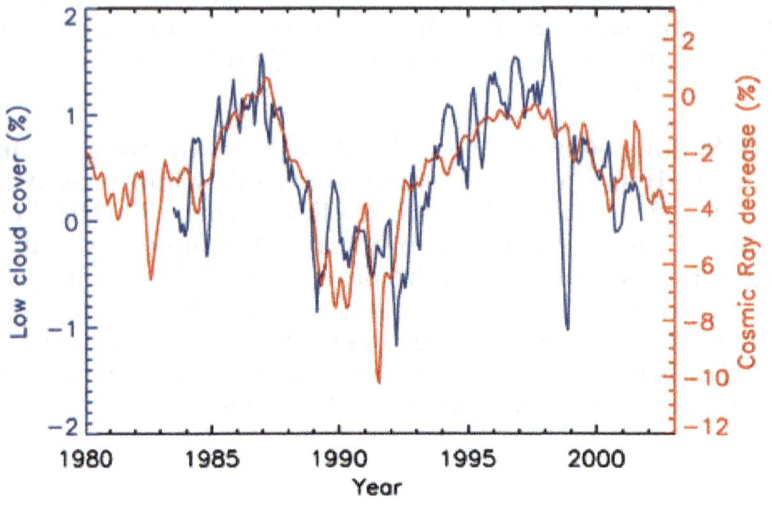

Correlation of cosmic rays and Earth's cloud cover.

Svensmark and his colleagues hypothesized that cosmic rays entering the atmosphere condense around organic molecules produced by trees to form the water droplets making up clouds. This causes more clouds to form, cooling Earth's surface.

During the low part of the sun's 11-year cycle when there is less solar wind, more cosmic rays are able to enter the earth's atmosphere and it is cloudier. During the 22-year period between 1983 and 2005, the correlation between cosmic rays and increased cloud cover was found to be 92%: extremely close. This has also been the case during the 400-year history of sunspot observation. When there are fewer sunspots — and thus more cosmic rays reaching Earth — there is a higher percentage of low clouds, which reflect sunlight and cool the earth.[15]

CERN laboratory in Switzerland, which studies collisions between fast-moving atomic-sized particles, tested this theory and verified it. You will never hear this from the media. Another fact the major media haven't reported is that when average temperatures on Earth were cooler, as they were in the 1970s, other planets in the Solar System, such as Venus and Mars, also had lower average temperatures. So solar activity is a factor driving natural climate change throughout the Solar System.

In 2017 Ned Nicolov and Karl Zeller published a paper entitled, "New Insights on the Physical Nature of the Atmospheric Greenhouse Effect Deduced from an Empirical Planetary Temperature Model," This report details the discovery that a comparison of the difference in the mean temperatures of rocky planets and moons: Earth, Mars, Venus, Mercury, Earth's moon, Luna, and Neptune's moon, Triton, with and without an atmosphere produces the same relationship. The temperature difference depends on the atmospheric pressures at the surface of celestial bodies with atmospheres. They call this difference the "atmospheric thermal effect" (ATE), rather than "greenhouse effect", since Earth is not a greenhouse. Venus has a thick atmosphere and very high pressure at its surface, which explains why the temperature of Venus is so high. Mars and Triton have thin atmospheres, and so there is a much smaller difference between their temperatures with and without an atmosphere. Using data from NASA, they discovered that Earth's temperature is 45% higher than it would be without an atmosphere.

In doing this they were able to calculate Earth's average temperature to within 1 degree Celsius. This study demonstrates that the temperature of a planet is really a function of its surface atmospheric pressure—not atmospheric composition. This coincides with the fact that when altitude increases, pressure decreases and so does temperature—called adiabatic cooling. This conclusion is a direct refutation of the CO_2 greenhouse hypothesis and, if correct, proves that the percentage of carbon dioxide—or any other gas—in the atmosphere has no bearing on the temperature.

Not all solar cycles are the same. In strong cycles there are many large sunspots, and in others hardly any. Solar cycles #22 and #23 were very active, which probably caused the 20 years of slight warming near the end of the 20th century. At the end of solar cycle #24, however, the sun was in a period of very low solar activity with no sunspots over a period of several days. The solar minimum between solar cycle #23 and #24 saw 260 spotless days in 2009 — the most days without sunspots in a single year since 1913.[16]

During the Little Ice Age, a prolonged period of low sunspot activity known as the Maunder Minimum preceded an era so cold that between 1645 and 1715 the Thames River in London froze up in the winter, and people regularly ice skated on it.[17] This never happens now. In the current solar cycle, #25, the sun is undergoing unusually high activity. This is probably one reason why temperatures have warmed a bit in the last few years.

The sun goes through other cycles besides the 11-year cycle. Otto-Carl Weiss and his colleagues examined European temperature data for the past two hundred fifty years using Fourier Transform spectral analysis. Their measurements indicated that, rather than a constant increase in temperature since the beginning of the Industrial Revolution, the climate had warmed and cooled periodically in ~200-year De Vries solar cycles. His team also concluded that climate change is natural with no influence from carbon dioxide.[18]

Another theory of the natural cause of climate change was introduced by Dr. Don Easterbrook, who conducted an extensive investigation of all warm/cold periods, not just full-scale Ice Ages and interglacial warm periods but also short-term warming and cooling periods.

He states: "For each warm/cold period, I looked at temperature, sunspot numbers (SSN), total solar irradiance (TSI), and radiocarbon (14C) and beryllium-10 production rates as indicators of cosmic ray intensity. The data are truly remarkable. Every cold period, regardless of the magnitude of the warmth or cooling, was characterized by low sunspot numbers, low TSI, and high production rates of 14C and 10Be, indicating high cosmic ray intensity. Every warm period was characterized by high sunspot numbers, high TSI, and low production rates of 14C and 10Be, indicating low cosmic ray intensity."

Dr. Easterbrook's study supports the conclusions of the Svensmark Clouds study, but does not depend on them. It stands alone on its own merits. The results of this study were summarized in the book "*Solar Magnet Cause of Climate Changes and the Origin of the Ice Ages*".[19]

Patterns of ocean currents are also a large factor in the warming and cooling of the climate. The Pacific Decadal Oscillation (PDO) is one such current. It reverses every 20 to 30 years.[20] When warmer water from the south Pacific moves through the Bering Strait into the Arctic Ocean, the Arctic is warmed and ice melts. The Pacific Decadal Oscillation, largely influencing the climate of the United States, switched to this pattern in 1910, after which warming started, and again in 1977, after which the warming pattern repeated. It is amazing how these two dates line up with the beginning of each warming period occurring in the 20th-century.[21]

It's true that the percentage of carbon dioxide, making up only about **four hundredths of one percent** (0.0420%) of the atmosphere, has been rising at a steady pace since the mid 20th century. It is also a fact that during the 1960s and 1970s, while the percentage of carbon dioxide was increasing, the earth was cooling. So who could then make the assumption that increasing the amount of carbon dioxide would cause temperatures to rise?

Computer Models

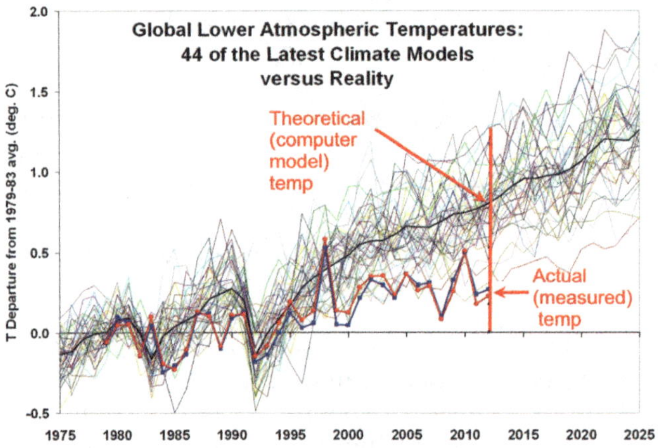

Computer model predictions are inaccurate.

Predictions of the amount of future warming on which the United Nations Intergovernmental Panel on Climate Change (IPCC) bases its claims hinge on data obtained by approximately 100 computer models. The problem with models is that if you don't correctly program into the computer everything, including the tiniest detail, involved with weather and climate, you cannot expect a valid result. Weather patterns are very complicated, and climatologists admit that many of these factors are not well understood. This is why predictions of future temperatures made by computer models have been so far off base. As shown on this graph, the warming they've predicted has been about three to four times as much as has actually occurred in the forty years since they started making these predictions. To reduce the amount of error, the people putting information into these models have periodically tweaked, or tuned them to conform to past records. In spite of this, they are still off by a great deal, as shown by this graph.

Dr. John Christy, University of Alabama professor and Alabama State Climatologist, discovered that when the model warms 1K, it allows 1.4 watts of heat to escape. But when the temperature rises 1K in the real world, 2.6 watts are released. Thus, the model derives an unrealistic amount of warming. Also, computer models are programmed to assume that carbon dioxide is the largest climate driver. In science we measure and record. Computer models do not measure anything. The results obtained by these models depend on what was programmed into them.

Dr. Christy also noted, "As such, the average of the models is untruthful in representing the recent decades of climate variation and change and thus would be inappropriate for use in predicting future changes in the climate or related policy decisions."[22]

The University of Alabama maintains the most up-to-date, accurate database of temperature information collected by satellites. This database reports 0.136 degrees C. of warming per decade, which amounts to a 1.4K temperature increase per century.

True empirical science is based on observations and measurements, using our five senses and instruments, of what is happening in the real world — not computer models. In order to study weather and climate, temperature, rainfall, wind speed and direction, air pressure and sea ice extent must be carefully measured, monitored, and recorded over a significant period of time.

The pieces of information gained in these measurements and tests are called data (a plural word). Data are carefully analyzed and used to draw conclusions and make predictions. This procedure is known as the scientific method. If the results or predictions of a hypothesis (proposed explanation) turn out to be incorrect, then it follows that either the hypothesis or the way of testing it is wrong. If the results conflict with the hypothesis, then it must be scrapped or changed. There's nothing wrong with changing your theory about an issue when not borne out by the data or when new data come to light.

After a report of a scientific study is carefully written by the researchers, they must submit it, along with their data, to a scientific journal for publication. This important process gives other scientists in that field (called peers) the opportunity to read and evaluate the study's conclusions and methods. Independent scientists can repeat the investigation, which should yield a similar result. This essential step of the scientific method is called peer review.

Because observations and experiments must be repeated over and over again, scientific investigation is difficult and time consuming. It could take years to complete a study and more years for other scientists working in the same field to verify the results. Only then can a hypothesis be considered a "theory". Established theories are also questioned and sometimes overturned. It is the very nature of science to be skeptical. While computer models could be helpful at times, they are no substitute for empirical science.

In 2012 forty nine former NASA scientists, including seven astronauts, sent a letter to NASA expressing their dismay that this agency — specifically the Goddard Institute for Space Studies (GISS) — was drawing conclusions based on scientifically flawed climate models that had failed to make accurate predictions and were ignoring proper scientific methods of investigation.[23] Some scientists have expressed their concern that when the truth about this deception does come out in the open, many people who once believed in the integrity of science will no longer trust science for answers. This is already happening in the health field, since it is now obvious that scientists lied to us about the origins of COVID-19. The treatments for this disease, such as Ivermectin and Vitamin D, were hidden from the public through censorship, causing the deaths of millions of people.

For decades scientists have been handsomely paid with government grants to perpetuate a human-generated crisis. If they stop doing this, their grants will be discontinued. In other words, they'll be out of a job. No wonder they've come to these startling and incorrect conclusions.

Changing Climate Data

As unbelievable as it may seem, agencies in the United States, Australia, New Zealand and other nations have actually **changed their own temperature records**. Many climate scientists have been surprised and shocked to see later editions of graphs published by the National Oceanic and Atmospheric Administration (NOAA) that had been changed from their previous editions.

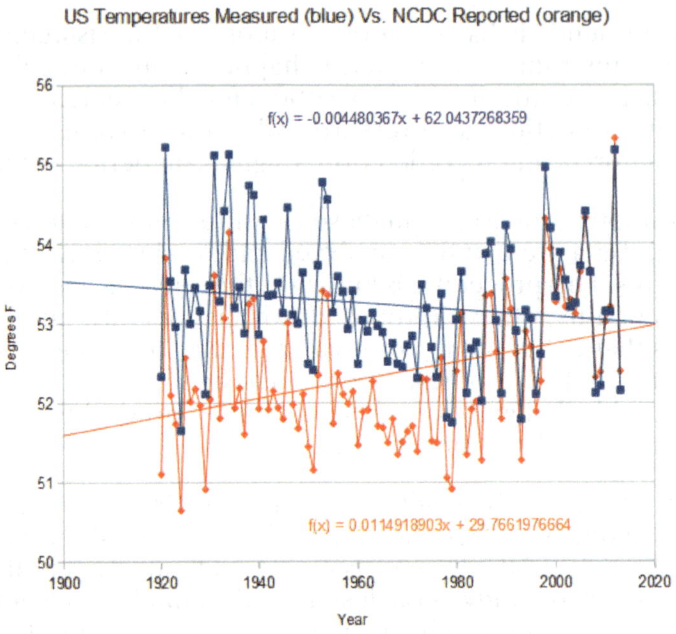

The National Oceanic and Atmospheric Administration (NOAA) and NASA actually lowered recorded temperatures from the early 20th century and raised more recent temperatures, making it appear that temperatures have risen. Just as in the novel *1984*, they actually changed history. This is not a conspiracy theory but an easily-provable fact, determined by comparing previously published graphs with the newer editions of the same graphs published later by these agencies after they had changed the data.

The blue line in the above graph shows the actual measured temperatures downloaded by scientists from official government websites. The orange line representing altered data was shown on the graph ultimately released to the public. As you can see, the temperatures from the early part of the 1900's have been lowered from what they actually were. The evidence points to the fact that in the 1920s and 30s temperatures were similar to today's and even warmer.

Tony Heller (RealClimateScience.com), who has carefully researched this matter, concluded that NOAA's "adjustments" actually represent the entire 'warming trend'".[24] This is worse than lying with statistics. It is entirely made up. Since 1976 the percentage of stations with estimated (false) data has grown from about 5% to 50%. Meteorologist John Shewchuk studied the data and verified that NASA and NOAA changed historical data, cooling older temperatures and warming recent ones.[25]

These agencies don't deny that they've cooked the data. With a flurry of double talk they admit that they've made "adjustments" to it. "In the year 2000 NASA charts showed about 0.5°C. of warming prior to 2000. But in 2023, the same graphs showed nearly 1.5°C warming prior to the year 2000".[26] Temperature data were also fudged under the direction of the Climate Research Unit (CRU) in Britain and also in Australia and New Zealand. This blatant fraud being committed right before our eyes goes against every principle of science.

If the "US Heat Waves" graph shown below had been superimposed on the climate record starting with the year 1900, it would have come out looking like the correct graph below it. Notice how much more severe the heat waves were in the 1930s and 40s.

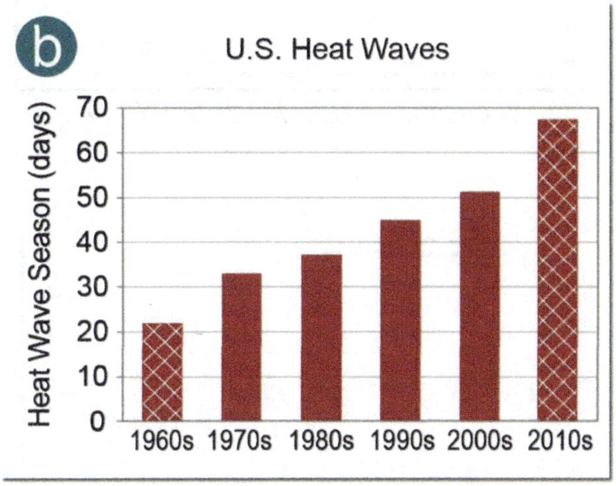

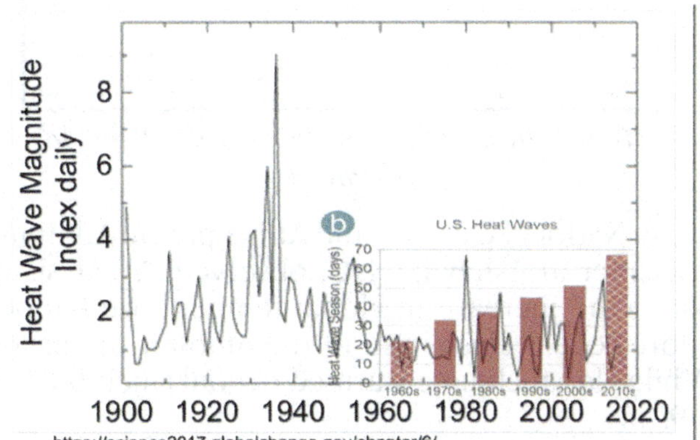

https://science2017.globalchange.gov/chapter/6/
https://nca2018.globalchange.gov/downloads/NCA4_2018_FullReport.pdf

Why did the graph released in 2018 show the 1960s as the starting point? Because the decade of the 1960s was the coolest of the past century. "The 2018 report released to the public cherry-picked the only start date in the graph which would create the appearance of a warming trend."[27] They simply acted as if the records previous to the 1960s didn't exist. This is one way of lying with statistics and is fraud of the highest order.

The two graphs below depict measured temperatures from 1996 to 2016. The 2016 edition shows the temperatures as they were measured. The 2020 edition of the same graph reflects temperatures that have been modified or "adjusted". The modified graph shows a steep rate of warming. These are not small changes.

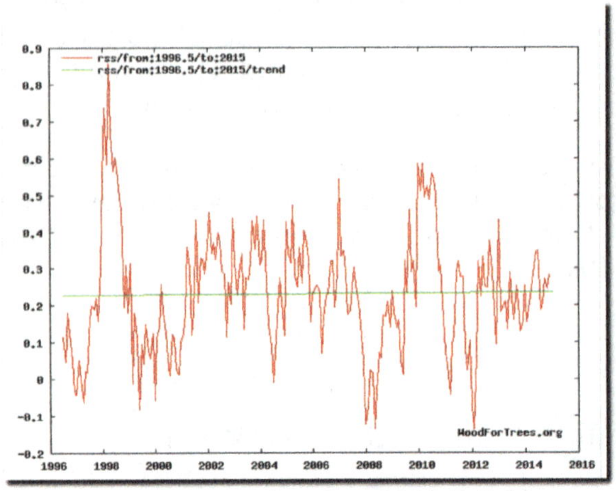

Temperature Graph - 2016 Edition

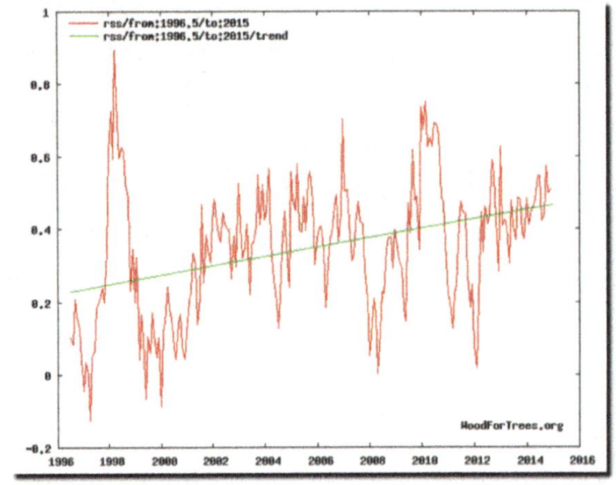

2020 Edition of the Same Graph (After Adjustments)

Thomas Karl, of the National Oceanic and Atmospheric Administration's National Climate Data Center in Asheville, NC, observed, "Analysis of warming since 1881 shows most of the increase in global temperature happened before 1919 — before the more recent sharp rise in the amount of carbon dioxide in the atmosphere... While global climate warmed overall since 1881, it actually cooled from 1921 to 1979."

In 2009 Dr. Alan Carlin, an employee of the Environmental Protection Agency (EPA), was asked to review the Draft Technical Support Document for Endangerment Analysis for Greenhouse Gas Emissions under the Clean Air Act. In an 85-page paper he made nineteen recommendations for revisions to that document. He and Dr. John Davidson criticized the EPA for accepting the Intergovernmental Panel on Climate Change's (IPCC's) mistaken conclusions and failing to do their own scientific research. Some of their conclusions were:

> "Global temperatures have declined extending the current downtrend to years. At the same time atmospheric CO_2 levels have continued to increase and CO_2 emissions have accelerated...The consensus on past, present and future Atlantic hurricane behavior has

> changed...Now the consensus is much more neutral, arguing that future Atlantic tropical cyclones will be little different than those of the past...The idea that warming temperatures will cause Greenland to rapidly shed its ice has been greatly diminished by new results indicating little evidence for the operation of such processes...Given the downward trend in temperatures since 1998 (which some think will continue until at least 2030), there is no particular reason to rush into decisions based upon a scientific hypothesis that does not appear to explain most of the available data."[28]

A senior EPA official, Al McGartland, quickly put the lid on this paper, admonishing its writers in an email, "The administrator and the [Obama] Administration have decided to move forward [with a cap-and-trade bill] . . .and your comments do not help the legal or policy case for this decision."

In the year 1911 a massive heat wave killed tens of thousands of people in Europe and the United States. This is verified by newspaper articles from 1911. In France alone 41,000 people died.[29] In Nashua, New Hampshire, the temperature reached 106° F, and the temperature reached 112° F in Maine. Thousands died in New England. Yet strangely, NASA's graph shows 1911 being a cold year. 1910, 1911 and 1912 were exceptionally hot years. Rapid melting of glaciers and calving of icebergs was reported. A collision with one of these icebergs at a latitude close to that of New York sank the Titanic on April 12, 1912.

A May 5, 1912, article in the *New York Times* reads, "An unprecedentedly warm winter in the entire Arctic is believed to be the cause of the vast number of icebergs in the north Atlantic Ocean during the present season and for the low latitudes which many of them have reached. . .Last year, according to officials at the Hydrographic Office, the thermometer reached 94 degrees in midsummer at Irigtut, on the west coast of Greenland." Yet, this graph by NOAA shows 1912 to be an exceptionally **cold** year, demonstrating the extent to which temperature records have been manipulated. This graph also indicates 1930 as being a cold year even though glaciers were melting at that time.

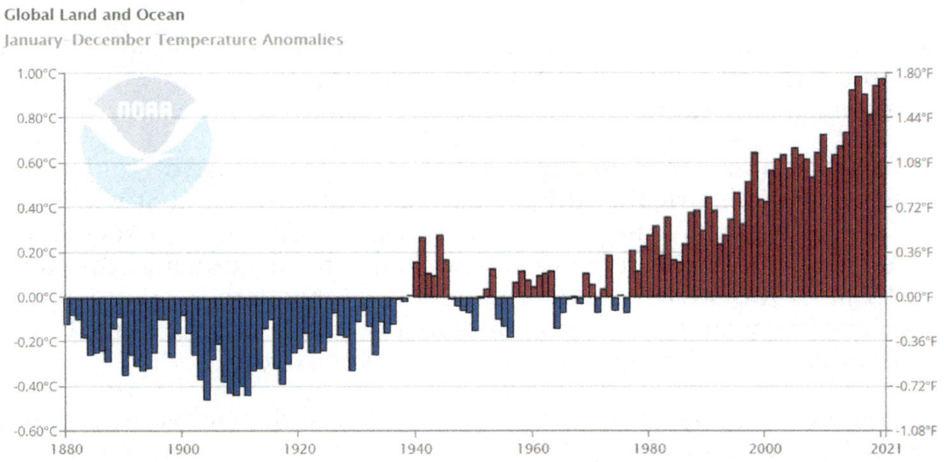

From *The Sydney Morning Herald*, January 13, 1930: "One of the riddles which is puzzling geologists all over the world is the continuous retreat of the ice glaciers. In his presidential address to the Science Congress, Professor R. Speight noted that the steady retreat of the glaciers in New Zealand had been observed over the previous 70 years. Photographs taken in 1896 and 1925 showed that several glaciers had retreated distances varying from 100 yards to half a mile during that time.[30] Melting of glaciers was a world-wide phenomenon, with similar reports in Scandinavia, the United States, Switzerland and Iceland. In Alaska glaciers had been retreating for the previous two centuries at the rate of 50–100 feet per year. Even Antarctica had shown signs of melting. [31] Is that what we're seeing on this graph?

The IPCC, CRU, NASA and NOAA are guilty of, and have admitted to:

1. Basing findings on computer data, not the real world,
2. Guesswork involving the Urban Heat Island effect,
3. Manipulation of temperature records to suit their agenda, and
4. Making up data for nonexistent (ghost) weather stations.

Three Equations Debunking the AGW Narrative

Recently Christopher Monckton, a former advisor to British Prime Minister Margaret Thatcher and renowned mathematician published in peer-reviewed climate journals, presented three equations that he believes will destroy the climate change alarmist narrative.[32]

The first equation corrects the equation climate scientists have used to calculate the amount of warming we'd have if the percentage of CO_2 (0.04%) doubled. Lord Monckton explains that the equation climatologists have been using is incorrect, because it fails to start with the baseline: Earth's temperature **without** greenhouse gases. Without the greenhouse effect, Earth's average temperature would only be about 260K (-13 degrees C). This number must be in the equation as a baseline signal. The corrected equation is:

260K (original temperature) + 29K (total greenhouse effect) ÷ 260K + 9K (direct greenhouse warming) = 1.1

This number turns out to be only about one third of the result obtained when the baseline signal is not used in the equation. In other words, this would amount to much less warming.

The second equation centers on the fact that as the percentage of "renewable" energy feeding an electrical grid increases, a point is reached at which the energy input at various times — mainly in the afternoon — exceeds the current demand, and the energy is thus wasted. This is known as the "duck curve". To meet the 50% goal governments have mandated, every wind or solar power plant would have to be overbuilt to a ridiculous extent. In this case renewable energy becomes prohibitively expensive.

Douglas Pollock, a Chilean engineer who discovered this phenomenon, was prevented from warning Latin American policymakers about this serious problem. This problem has already surfaced in California and in seven countries. [33]Douglas Pollock's equation is:

E (excess power) = N (nameplate capacity) – D (hourly demand)

The excess power is wasted, because hardly any of it can be saved in batteries. This is because battery technology has failed to keep up with the requirements for such a system. So these power companies will never be able to reach 50%. It makes no sense to mandate 50% unless the policy makers setting these requirements are hell bent on destroying economies.

The third equation involves the amount of warming that would be prevented by 2050 if the entire world produced no net CO_2 emissions from now until that time. The equation is:

Warming prevented = 0.45 units of influence abated x 1.8 TCR (transient doubled CO_2 response) ÷ 3.93 ERF (effective radiative forcing) units = 0.2K.

Therefore, if somehow all human-caused CO_2 emissions were to immediately cease, by the year 2050 just 0.2 of a degree of warming would be averted. This would cost many trillions of dollars, even if possible, and no one would even notice 0.2 degrees. Christopher Monckton pointed out that "The UK national grid has estimated $3.6 trillion as the cost of re-engineering the grid to meet the net-zero target."

How Real Science is Done

1. Make careful observations and ask questions.
2. Formulate a hypothesis (guess) as to the cause of the observed phenomena.
3. Test, measure and carefully record observed data.
4. Analyze the results of your observations and measurements.
5. Be ready to change the hypothesis if it doesn't fit the data.

How Climate Pseudoscience is Done

1. Start with your conclusion (your agenda): "Mankind's CO2 emissions are making the earth boiling hot."
2. Think of a possible hypothesis of your conclusion: "The percentage of carbon dioxide is rising. This caused the temperature increase."
3. Change the data to fit your hypothesis: "Just ignore the fact that it was warmer in the decades before 1960."
4. Personally attack or censor scientists who disagree with you. Call them "science deniers" or "oil shills". Threaten them with prison.

These are facts, not "conspiracy theories". How can anyone take the alarmists' climate change claims seriously when they have been found guilty of such a grievous departure from the truth and the violation of correct science procedure?

---

1. Ben Bikman, PhD
2. WeatherAction.com

3. Robert Zubrin, Merchants of Despair, New York: Encounter Books, 2012, p. 227
4. Ball, op. cit.
5. Larry Bell, "The New York Times' Global Warming Hysteria Ignores 17 Years Of Flat Global Temperatures," Forbes Magazine, Aug. 21,2013
6. Anthony Watts, "Bombshell – NOAA Whistleblower says Karl et al. "Pausebuster" Paper Was Hyped, Broke Procedures," WattsUpWithThat.com, Feb. 4, 2017
7. Tony Heller, YouTube video, "Coldest October-April In The US In Over A Century," May 2, 2019
8. Topher Field, "50 to 1David Evans Interview," YouTube, Sept. 1, 2013
9. Knappenberger, https://www.masterresource.org/.../climate-sensitivity-estimates-heading-down-way-down
10. Joannenova.com.au/2009/12/the-one-flaw-that-wipes-out-the-crisis/
11. David Evans, The Australian, July 18, 2008
12. David Evans, "Is the Western Climate Establishment Corrupt?", Nov. 8, 2010, http://science and publicpolicy.org/images/ stories/papers/originals/western_climate_establishment corrupt.pdf
13. Ball, op. cit., p. 8
14. Mary Ann Gebauer, YouTube Channel, interview on Sept. 2, 2025
15. Larry Vardiman, A New Theory of Climate Change, 11/1/08 https://www.icr.org,
16. http://www.climatedepot.com/2018/05/23/are-we-headed-towards-another-deep-solar-minimum/
17. Mark Prigg, "Is a Mini Ice Age On the Way?", DailyMail.uk, July 10, 2015
18. Lecture by Prof. Carl-Otto Weiss, Advisor to the European Institute for Climate and Energy, "International Conference of the Schiller Institute", Paris, France; June 14, 2015
19. Dr. Don Easterbrook, from abstract, July 9, 2019
20. https://sealevel.jpl.nasa.gov/data/el-nino-la-nina-watch-and-pdo/pacific-decadal-oscillation-pdo/
21. Dr. Fred Goldberg lecture March 15, 2016 (YouTube video)
22. John Christy, House Comm. Science, Space and Technology, (March 29, 2017) p.13
23. Cfact, "Astronauts and scientists send letter to NASA: Stop global warming advocacy," www.cfact.org, April 10, 2012
24. Tony Heller, "Climate Scientists Rewriting The Past," www.realclimatescience.com/climate-scientists-rewriting-the-past/
25. Katie Spence, *Climate Change Dispatch*, Jan. 29, 2024.
26. Ibid.
27. Ibid.

28. "EPA Dept. Director Kills Report finding Global Warming is an invalid hypothesis", July 7, 2009, climetechsolutions.com

29. https://climatechangedispatch.com/france-1911-heatwave-41000-deaths/

30. Speech to Royal Society of New Zealand, *Otego Daily Times*, May 18, 1935, PapersPast

31. "Arctic is warmer; glaciers retreat. . .Spitzbergen once closed to ships from October to May, now approachable through winter," *Evening Star*, Washington, DC., Feb. 18, 1925

32. Christopher Monckton, "Three Damning Equations To Defeat the Global Warming Zealots," American Thinker, May 12, 2023

33. Germany, Spain, Ireland, the United Kingdom, Italy, Australia and Chile

Chapter 6

False Climate Change Claims

> "The great enemy of the truth is very often not the lie, deliberate, contrived and dishonest, but the myth, persistent, persuasive and unrealistic." —John F. Kennedy

"Polar bears face extinction due to the melting Arctic."

The headline of an undated article on the World Wildlife Fund website reads, "Polar Bear Population Decline a Wake-up Call for Climate Change Action."[1] In reading this article we learn that it is the polar bear population in the Beaufort Sea — not the world's polar bear population — that has declined, another example of misleading and twisting information.

Polar Bears: courtesy of lakeshoresolar.com

In contrast, a 2011 article in International Business Times stated, "The US Fish and Wildlife Service estimates that the polar bear population is currently at 20,000 to 25,000 bears, up from as low as 5,000-10,000 bears in the 1950s and 1960s.[2]Another estimate of the current polar bear population is between 22,000 and 31,000 bears. [3]This number is verified by the IUCN/SSC Polar Bear Specialist Group.[4] Because of strict hunting regulations, the number of polar bears has been increasing for many years.

Dr. Susan Crockford, one of the world's premier experts on polar bears, reported:

> "Polar bears turned out to be more flexible and resilient than predicted and many subpopulations are better off than before...There is no existential emergency for polar bears or any other Arctic sea mammals due to declining summer sea ice, despite continued messages of doom from remorseless experts."[5]

"97% of scientists say that human-caused climate change is a serious threat."

We are repeatedly told that a 97% consensus [agreement] of scientists say that Earth's climate is experiencing dangerous warming caused by human activities. This erroneous and totally misleading figure was obtained from four different studies, all of which were flawed.

The first was by Naomi Oreskes. The topics of many articles counted in her study failed to address the causes of climate change, but — assuming that IPCC conclusions were correct — instead focused on the influence of climate change on the incidence of influenza, the life cycle of frogs, etc. Before publishing Ms Oreskes' article, the journal editor failed to check either her methodology or her sources.

In a 2009 study Maggie Zimmerman, a graduate student, sent a 2-minute online survey to 10,257 employees of schools and government research agencies. This left out over 10,000 geologists, physicists, meteorologists and astronomers who were studying the issue. Although she received 3,146 responses, her 98% figure was based on only 79 responses that she had hand picked. This was certainly not a representative sample.

In 2010 another college student, William Anderegg, identified 908 scientists who had written the most papers about global warming. He considered these to be most qualified to hold an opinion on climate change. "The 50 most prolific alarmists were published an average of 408 times each, versus only 89 times each for the skeptics." Hundreds of the studies he considered had several — as many as twelve — different authors. This misinformation weighted the results against the skeptics.[6] Anderegg failed to consider that the quantity of published studies doesn't equal correctness.[7] Scientists whose studies reach conclusions agreeing with the official government position on climate change are rewarded with lucrative grants. In contrast, those scientists who study the natural causes of climate change must scramble for financing. This is one reason why these scientists aren't publishing as many peer-reviewed papers.

In the fourth study John Cook, a professional cartoonist, issued a report to the Global Warming Policy Foundation stating that 97.1% of the 11,944 scientific papers he examined "explicitly or implicitly suggested that human activity was responsible for some warming."[8] This claim has been debunked by several other studies, including one by Legates which found that only 41 of these 11,944 papers (or 0.3%) stated that most of the warming since 1950 was man made.[9]

Cook himself admitted that about two-thirds of the abstracts [summaries] he looked at expressed no opinion about manmade global warming. These were excluded from the results, revealing the dishonest character of the study. Laughably, Cook had told the publisher of his study its results (97% consensus) **before** he started the research.[10] Regarding COVID-19, one person quipped, "90% of scientists agree with the people who fund them." This is also obviously true of climate scientists. It's human nature.

But science doesn't operate by consensus. To adhere to the scientific method, scientists are obligated to question all new ideas. In that case, even if this 97% figure was correct (it isn't), it would be meaningless. When Galileo challenged the scientific consensus that the earth was the center of the universe, he was met by an extremely hostile reaction.

On the other hand, in the 2007 Global Warming Petition project 31,487 scientists, over 9,000 of whom had PhDs, signed the following declaration:

> "There is no convincing scientific evidence that human release of carbon dioxide, methane or other greenhouse gases is causing, or will, in the foreseeable future, cause catastrophic heating of the Earth's atmosphere and disruption of the Earth's climate. Moreover, there is substantial scientific evidence that increases in atmospheric carbon dioxide produce many beneficial effects upon the natural plant and animal environments of the earth."[11]

A prominent scientist, Hal Lewis, resigned from the American Physical Society (APS) in 2010 after noting that this society had been taken over by unscientific global warming hysteria. In his letter of resignation, he stated: "The global warming scam, with the trillions of dollars driving it, that has corrupted so many scientists, has carried APS before it like a rogue wave. It is the greatest and most successful fraud I have seen in my long life as a physicist."[12]

Many scientists have been afraid to disagree with those in power out of fear that their careers will be destroyed. Since the global warming hypothesis surfaced about 40 years ago, many meteorologists and climatologists have waited until they retired to speak up. Others became skeptical after more information surfaced. Just a few of the top scientists who are now standing up and declaring that this is wrong are: Dr. David Evans, Dr. Joanna Simpson, Dr. Judith Curry, Dr. Alrude Allegre, Dr. Tad Murty, Dr. Chris de Freitas, Dr. Kiminori Itoh, and Dr. Andrei Kapitsa.[13]You need to look up the names of these scientists to learn why they've changed their stance.

Scientists choosing to investigate natural rather than man-made causes of warming are denied research grants. This is true obviously because government officials, lusting for carbon taxes, are interested only in those reports that can show in some way that humans are causing this "problem". For example, if a scientist can demonstrate that a species of butterfly has migrated north or is in danger of extinction, attributing this to human-caused climate change will guarantee them lucrative grants that are perpetually renewed. Scientists are human beings, and they don't want to have their careers cut short by losing their government grants. They know these grants will be renewed as long as there is a "problem".

The voices of dissenting scientists are drowned out by the tsunami of voices of climate hysteria. On Nov. 29, 2012, 125 scientists wrote an open letter to the United Nations [14]declaring that there is no climate emergency. This was ignored. On Sept. 23, 2019, 500 prominent scientists sent a letter to this international body, stating:

> "The general circulation models of climate on which international policy is at present founded are unfit for their purpose. Therefore, it is cruel as well as imprudent to advocate the squandering of trillions of dollars on the basis of results from such immature models. Current climate policies pointlessly and grievously undermine the economic system, putting lives at risk in countries denied access to affordable, reliable electrical energy."

"Ice in the polar regions is melting as never before."

The first scene of a 2004 science fiction movie, "The Day After Tomorrow", showed a coastal portion of the West Antarctic ice sheet crashing into the ocean. But recently, after detecting unexplained seismic activity in that region, geologists from Washington University in St. Louis accidentally discovered the cause of that warming: a chain of about 90 active volcanoes hidden under this glacier.[15]

According to NASA, the total amount of ice in Antarctica has been increasing for several decades.[16] The Antarctic glacier, containing about 90% of the earth's fresh water, is exceptionally stable. It is almost five kilometers (three miles) thick in the center. NASA stated, "The Antarctic ice sheet showed a net gain of 112 billion tons of ice per year from 1992 to 2001."[17]

But recently, despite these earlier gains, it was falsely reported that there has been a steep downturn in Antarctic ice. "Scientists wonder if it's 'the beginning of the end'"[18]With their usual flair for exaggeration, the doomsayers claim that the Antarctic is heating up out of control. But satellite photos of the continent show more sea ice around Antarctica than existed in a 1976 photo in *National Geographic.* Computer models predicted heating in Antarctica to be twice what it actually turned out to be. In 2020 *CNN* reported that Antarctica had the coldest temperatures on record. [19]This graph shows the temperature trend since 1980.

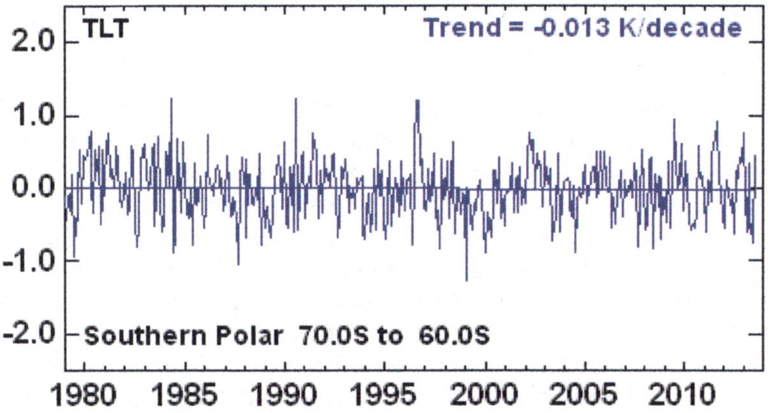

The average temperature in Antarctica is -58° F. To get the ice to melt, you'd have to raise that temperature about 90° to water's melting point of 32°F. That's a huge increase in temperature. There have been no gaps in the Antarctic ice core record for the last 15 million years.[20] The snow and ice at the South Pole has been accumulating so rapidly that the first two scientific observation stations had to be abandoned after being taken over by ice. The third station was constructed on stilts so that it could be raised as needed.[21]

The collapse of a 160-square-mile block of the Wilkins Ice Shelf in March, 2008, was conclusive proof for the alarmist media that the entire continent (14.2 million km^2 or 5.5 million $miles^2$) was melting. A headline in the *Los Angeles Times* was "Antarctica Collapse", which was described as the "rapid melt of the Wilkins Shelf."[22] Not to be outdone, *The Sydney Morning Herald* claimed that "The Ice Shelf Hangs by a Thread."[23] The headline in *Salon* magazine was "Bye-bye, Antarctica?"[24]This type of exaggeration can only be considered a lie, but not an innocent one. Joseph D'Aleo, the former Director of Meteorology for the Weather

Channel, compared this event to "an icicle falling from a snow and ice-covered roof."

But what about the Arctic? We've also been told that the Arctic ice is melting. Well, a plane that crash landed on the ice in Greenland in 1942 was recovered fifty years later in 1992 under 268 feet of ice. This demonstrates that Greenland's ice has been getting thicker, not thinner. It doesn't get warm enough in the summer to melt all the ice that formed the previous winter. The Greenland glacier is melting at the edges but getting thicker in the interior. Of course, every summer there is some melting of icebergs near the ocean. The following winter this melted ice is replaced by an equal amount of snowfall. That cycle has been occurring for thousands of years.

A good example of a half-truth so common in our media is the following statement: "Melting in the Arctic reached an all-time high in June. Ice has been disappearing at 29,000 square miles a day."[25] The media likes to show photos and videos of glacial icebergs calving off and falling into the sea. This turns out to be another good way of lying with statistics. Dr. Tim Ball explained what is really happening here: "This is near the average daily rate of melt in the brief Arctic summer, but few people know this is natural. Approximately ten million km^2 melts every summer in approximately 145 days."[26]About the same quantity of this melted ice — an area about the size of the continental US — is replaced the following winter.

Receiving direct sunlight, the tropics get much more energy from the sun than the polar regions. Twenty percent of the sun's heat impacts tropical oceans. This heat moves to the polar regions, where it is radiated back out to space. The IPCC doesn't acknowledge this.

National Geographic magazine predicted that all of the snow and ice in the Arctic would be melted by 2008. Canada.com predicted this event would happen between 2010 and 2015. A leading expert at NASA predicted that all the snow and ice in the Arctic would be gone by 2012. "The Arctic is screaming," declared Mark Serreze, a senior scientist at the National Snow and Ice Data Center in Boulder, Colorado. [27]The BBC put the year for disappearing Arctic ice at 2013. Nobel prize winner, Al Gore, predicted that all the ice would be gone in 2014. Being more conservative, James Hanson of NASA's Goddard Space Center named 2018 as the year for this occurrence. But, as you can see from this graph, in 2022 the average Arctic ice extent was near the top of the average number of square miles for the 30-year period of 1981 to 2010.

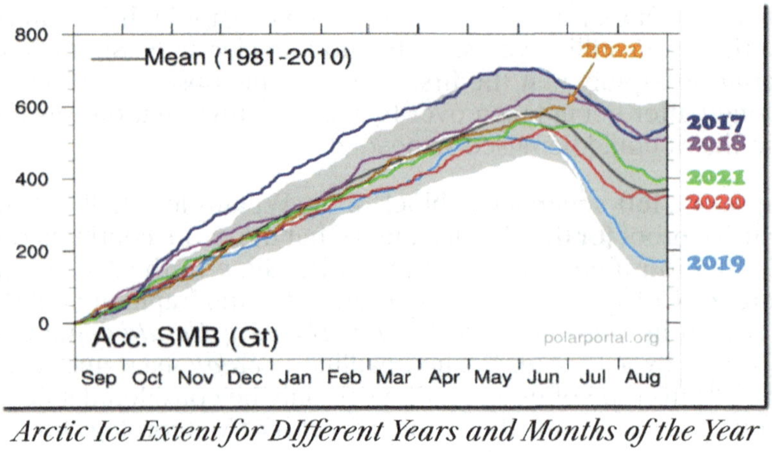

Arctic Ice Extent for DIfferent Years and Months of the Year

A large Greenland glacier, named Jakobshavn, which had been thinning out around 2012, started growing again in 2017, according to NASA's Oceans Melting Greenland (OMG) project. The study's lead author, Ala Khazendar, a NASA glaciologist on the OMG project, believes that the cooling of the North Atlantic due to the shift in the Atlantic Multidecadal Oscillation probably caused that glacier to rebound. Other NASA scientists have called this a "miracle".[28] The Atlantic Multidecadal Oscillation, a repeating 50 to 70-year cycle of changing ocean currents in the Atlantic Ocean, influences the temperatures of Greenland and other Arctic areas.

The North Pole is covered by sea ice averaging about two meters in thickness. There have been several warmer times in recorded history — as recently as 1937 — when ships were able to navigate through the Northwest Passage. This isn't possible at the present time. So how could 2024 have been the "warmest year in 125,000 years"?

For generations alarmists have been predicting that Arctic ice would soon disappear. Newspaper and magazine articles in the early 1900s also made that prediction.[29] From 1880 to 1915 glaciers advanced. Then, during the period from 1915 to 1945 they receded. They started advancing again from 1945 to 1977 when the earth was again cooling.

The red trend line in the following graph shows that the extent of ice in the Arctic has been almost perfectly constant during the period from 2006 to the present time. The blue line moving up and down reflects the yearly freezing and thawing in response to summer and winter temperatures. There was a slight reduction in the amount of ice in 2012, but it quickly rebounded. The MASIE-NH Daily Image Viewer, published by NOAA, shows the amount of snow and ice in this area at any particular time. It can be accessed by clicking the link: https://nsidc.org/data/m asie.

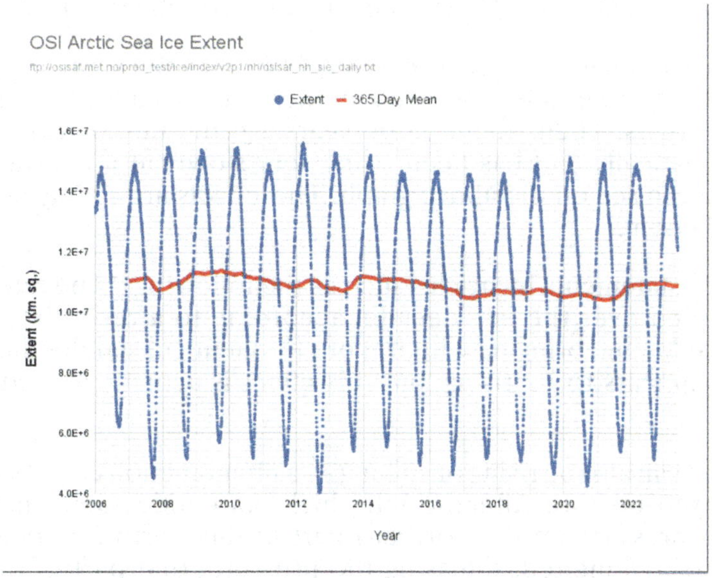

Recently a team of scientists from Thailand who studied over 30,000 satellite records concluded that Greenland has actually been cooling for the past two decades.[30] This is a surprise, since the World Economic Forum recently stated that the Greenland ice sheet is collapsing.

"Sea Level Rise is Accelerating."

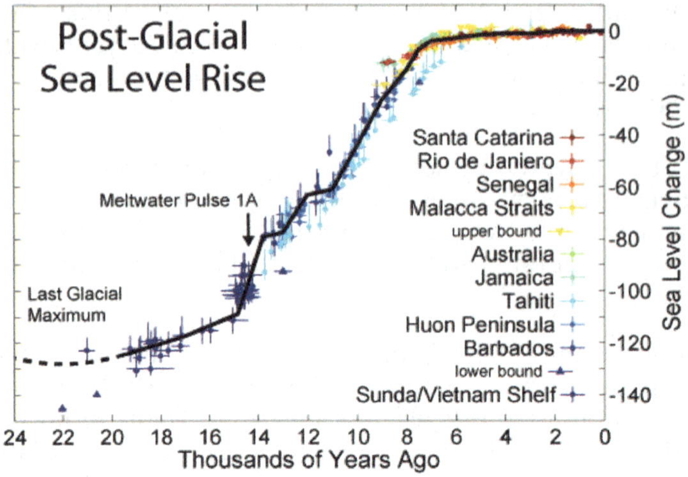

For the last 8,000 years the earth's sea level has been slowly rising at a constant rate. Nils Axel Mörner, a world-renowned expert on global sea level measurement, warned the UN Intergovernmental Panel on Climate Change (IPCC) that it was publishing false information that would inevitably be found to be incorrect. When they ignored him, he resigned.

Dr. Mörner has affirmed that Earth's climate is cooling and that ocean levels will decrease. He reports that the sea level has been rising at an average rate of just 1.1 mm — about the thickness of your fingernail — per year. This figure is confirmed by NOAA.[31] This sea level increase adds up to about 18 cm (7 inches) per century, not the 1-2 meters (39 -78 inches) per century incorrectly reported by the media.

The warming following the last glaciation of the current Pleistocene ice age was accompanied by a rapid ice melt and an increase in sea level over a period of about 7,000 years. Then, about 8,000 years ago the rate of sea level increase slowed dramatically and has been rising very slowly at a constant rate since then, as you can see on the above graph. These facts are simply ignored by the prophets of doom.

We often see statements in articles such as the following: "The rate of sea-level rise has tripled over the last decade, according to a recent study from the University of Miami, bringing with it more frequent coastal flooding."[32]The sea level there appears to be rising, but this effect is actually caused by the land sinking.

Because land can also be rising or sinking — a phenomenon called isostasy — sea level is tricky to measure. After the mile-thick Laurentide glacier that was putting tremendous pressure on the northern part of the North American continent 20,000 years ago melted, releasing the pressure, that part of the continent started rising, while the southern end of the continent, bordering the Gulf of Mexico, began to sink.[33] This includes the Miami area, which is why we've been told that sea level is increasing there. The Houston area is also sinking.

"Subsidence [sinking] is a pernicious, highly localized and often overlooked problem in comparison to global sea level rise, but it's a major factor that explains why water levels are rising in many parts of the eastern US."[34]Another cause of subsidence of these great coastal cities is the amount of water that has been removed from the water tables beneath these cities.

Sea level measurements are made both by averaging tidal gauge data and by satellite altimetry. Dr. Mörner complained that the satellite data was tampered with to incorrectly show a rise in sea level.[35] Another system called GRACE – made up of gravitational-anomaly satellites - computes the mass of the ocean, from which changes in sea-level can be calculated.

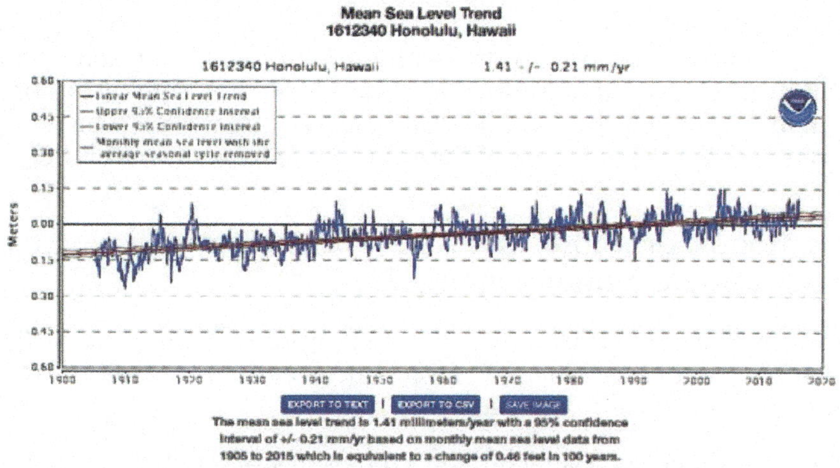

If the sea level was undergoing a significant increase due to CO_2 , the line in this typical tidal gauge graph would curve upward as the atmospheric percentage of carbon dioxide increased. But instead it is a straight line. If anything, sea level rise is slowing down slightly.

Many media reports have falsely claimed that, due to the rising sea level, the Maldives Islands in the Indian Ocean are being submerged underwater and that its people will thus have to be relocated. But Dr. Mörner stated, “Clear observational measurements in the field indicate that sea level is **not** rising in the Maldives, Bangladesh, Tuvalu, Vanuatu, or French Guyana.

“The president of Tuvalu continues to claim that his islands are being flooded. However, tidal-gauge data provide clear indications of stability over the last 30 years.[36] In Vanuatu, the tidal gauge also indicates a stable sea level over the last 14 years.”

The Beach at Tuvalu.

There was an earthquake near American Samoa in 2019. Since then that island has been sinking at the rate of 15 cm (6 inches) per decade. In a 2024 video UN Secretary General Antonio Guterres used this fact to claim that sea level is rising fast and will soon come to all coastal areas, another misrepresentation (lie) for which there is no excuse.

Dr. Mörner observed that there’s something “sick” in the IPCC’s methods. “In science we discuss. We don’t forbid or neglect. . .They are working with lobbyists. They blacklist those who disagree.”[37]

A study conducted by team of researchers led by Dr. John R.L. Allen, published in *Geophysical Research Letters* in 2018, used tide gauge data to challenge the prevailing narrative of sea level rise acceleration. The study found that there has been no acceleration in sea level rise for the last 100 years. This study computed the average rate of sea level rise at about 1.9 mm/year. This would amount to an increase of about 15 cm (6 inches) per century. So we've also been lied to about sea level rise.

On the last day of Barack Obama's presidency in January of 2017 NOAA issued a report that sea level could increase by 2.54 m (8 **ft)** by the end of this century. (They were off by 16 times.) Then Mr. Obama promptly bought a $19-million beachfront estate in Massachusetts and a multimillion dollar mansion on the beach in Hawaii, showing that he obviously wasn't concerned about any projected rise in sea level. Yet, despite the observational evidence — on which empirical science is based — the purveyors of doom continue to advance this lie.

"Storms, floods, droughts and forest fires are much worse now due to man-made climate change."

When the earth's temperatures leveled out after the El Nino event of 1998, the fear mongers seized on the idea of claiming that extreme weather events have been getting worse and more frequent.

Here again real science and statistics tell a different story. In fact, Bjorn Lomborg concluded that your risk of dying in a climate-related disaster has been reduced by 99% since 1920. This is because we have better systems in place to deal with these challenges. Dr. Roger Pielke, Jr., a mathematician, provided irrefutable statistical evidence clearly demonstrating that these claims are false. In sworn testimony before the US Senate Committee on Environment & Public Works, he laid out the following documented facts:

"It is misleading, and just plain incorrect," he said, "to claim that disasters associated with hurricanes, tornadoes, floods or droughts have increased on climate timescales either in the United States or globally. It is further incorrect to associate the increasing costs of disasters with the emission of greenhouse gases."

"Globally, weather-related losses [$] have not increased since 1990 as a proportion of GDP [they have actually decreased by about 25%], and insured catastrophe losses have not increased as a proportion of GDP since 1960."

"Hurricanes have not increased in the US in frequency, intensity or normalized damage since at least 1900. The same holds for tropical cyclones globally since at least 1970."

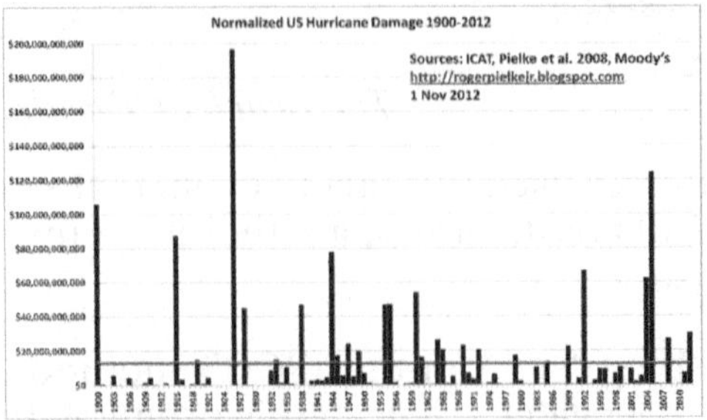

Cost of Hurricanes

As you can see from this list of the ten worst hurricanes in US history (#1 being the worst), the great majority occurred before the level of carbon dioxide in the atmosphere began to rise. The only recent ones were Hurricane Katrina in 2005 and Hurricane Ian in 2022.

1) 1900 Great Galveston Hurricane
2) 1928 Okeechobee Hurricane
3) 2005 Hurricane Katrina
4) 1893 Chenière Caminada Hurricane
5) 1893 Sea Islands Hurricane
6) 1881 Georgia/South Carolina
7) 1957 Hurricane Audrey
8) 1935 Great Labor Day Hurricane
9) 1856 Last Island Hurricane
10) 1926 Great Miami Hurricane[38]

Hurricane Ian, which hit Florida, North Carolina and South Carolina in late September of 2022 was not the 4th strongest hurricane to hit Florida, but the 10th strongest, according to meteorologist, Dr. Neil Frank. "An analysis of major hurricane landfalls by decade shows a significant downward trend – not the upward trend predicted by climate change activists. . .Today's landfalls do not compare to those in the middle of the last century. Florida was hit by seven major hurricanes in the 1940s. Six major hurricanes slashed the east coast in the 1950s. But not a single major hurricane made landfall in the entire United States from 2005 to 2017."[39]

2025 was predicted to be a strong hurricane season in the Atlantic. In fact, the 2025 hurricane season was well below average both in the number and intensity of storms. According to NOAA, from 1991 t0 2020 the average annual number of storms was 12.1; in 2025 it was 7. The average annual number of hurricanes is 6.4; in 2025 it was 2. The average annual number of strong hurricanes is 2.7; in 2025 it was zero. The Accumulated Cyclone Energy (ACE) index averages 91.4; in 2025 it was only 27.2. So here again their prediction was absolutely wrong. This again demonstrates how little climate alarmists understand about the climate.

Climate change alarmists frequently cite Hurricane Katrina and "superstorm" Sandy, as "proof of the effects of climate change." Katrina, which made landfall on Aug. 29, 2005, was a truly wicked hurricane, but it was not the worst hurricane ever to hit the United States. Much of the flooding damage was from levees that broke during the storm because they hadn't been properly maintained.[40]Sandy was just a category-1 hurricane. But it did a great deal of damage due to a bizarre set of circumstances. First, as it moved up the east coast of the US in late October of 2012, it slammed into an extremely cold air mass moving south from Canada. This cold air mass added a good deal of energy to the system.

Sandy, a Category 1 Hurricane

Sandy made landfall just south of Atlantic City, New Jersey, at the height of a spring tide, an unusually high tide that occurs when the sun and moon are lined up in such a way that they are both pulling in the same direction. "The storm surge — an additional 2.75 m (9 ft) or more of water piled up against the coast by furious winds and crashing ocean waves — was riding on top of this extra high tide."[41] Strangely, adjacent areas that were hardly impacted at all by this storm experienced mysterious power failures or otherwise went offline.

"Floods have not increased in the US in frequency, intensity or normalized damage since 1950. Flood losses as a percentage of US GDP have dropped by about 75% since 1940."

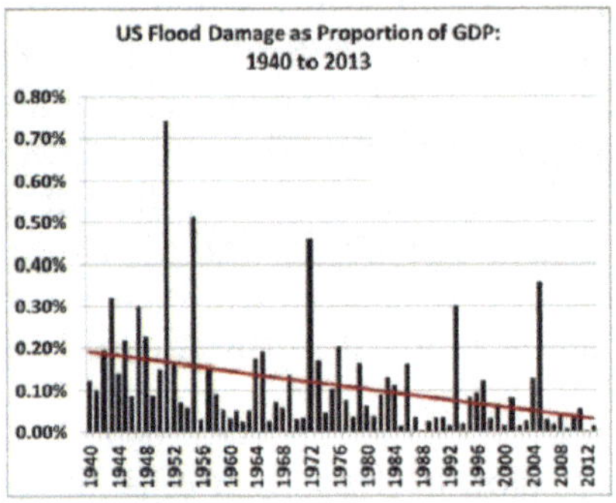

"Tornadoes have not increased in frequency, intensity or normalized damage, and there is some evidence to suggest that they have actually declined. Drought has for the most part, become shorter, less frequent, and covers a smaller part of the US over the last century. There has been a significant decrease in F5 tornadoes since 1950. You could say we are having an F5 drought. Alarmists want to lower the designation of F5 from the current 200mph to 190 mph. Globally, there has been little change in drought over the last century."[42]

These carefully compiled statistics were part of Dr. Pielke's sworn testimony before Congress. Because this information couldn't be disputed, global warming alarmists pulled out all stops to viciously attack his credentials, character and reputation. They claimed that he was "outside of the science community" and "on the fringe." The University of Colorado, where he is a professor, came to his defense.

It is also being claimed that the number of hectares/acres burned in forest fires has increased because of climate change. Gov. Gavin Newsom of California, recently spoke at the COP-30 conference in Brazil, where he was introduced as the "first climate president of the US." Evidently he plans to run for president in 2028. He gave an impassioned description of the terrible fire that burned the Palisades in Los Angeles, implying that it was caused by climate change. This fire took 25 lives, destroyed 5,000 structures and left thousands homeless. Evidence of the real causes points to a combination of arson, government negligence — with drained reservoirs and dry hydrants — and environmental mismanagement. Residents sued the city of Los Angeles, alleging that the Palisades Reservoir had been emptied for repairs and never refilled.

Regarding fire statistics, the facts speak for themselves. The Global Fire Emissions Database (GFED) which provides estimates of the burned areas around the globe, stated that from 1997 to 2016, 440 million hectares (1.1 billion acres) were burned each year. This number was lower than the total number of hectares burned annually during the twentieth century: 550 million hectares (1.4 billion acres).[43]

"Earth is heading for a mass extinction due to climate change."

A mass extinction event is characterized by the loss of a significant number of species of plants and animals: over 75% of them, within a relatively short period of time, 2.8 million years or less. In its history the earth has experienced five mass extinctions, the latest being the one that led to the extinction of the dinosaurs about 65 million years ago. The sixth, called the Anthropocene Extinction, is said to involve human-caused biodiversity loss happening at an unprecedented speed.

A 2015 study published in the journal *Science Advances* came to the following outrageous conclusions: "Modern extinction rates for vertebrates varied from 8 to 10 times higher than the background rate. . .Most extinctions have occurred in the last 114 years. . . Current extinction rates vastly exceed natural average background rates."[44] It is no wonder that the lapdog media ran with this story to frighten the daylights out of people, especially our young people. This undoubtedly prompted Greta Thunberg to exclaim, "We are in the beginning of a mass extinction, and all you can talk about is money and fairy tales of eternal economic growth. How dare you!"

However, a 2025 study has shone a bright light on this issue. This study, "Unpacking the Extinction Crisis: Rates, Patterns and Causes of Recent Extinctions in Plants and Animals," was published in the *Royal Society* journal in October of 2025. Like the one in 2015, this study used the data from the International Union for Conservation of Nature (IUCN) which has lists of extinct, as well as endangered and threatened extant species. They based their report on the last five hundred years since about 1500 AD.

The conclusions they reached were totally different than those of the previous study. These authors concluded, "Extinction rates have increased over the last five centuries, but generally declined in the last 100 years." [45]Another significant finding of this study was, "These past extinctions also do not support the idea that biodiversity loss is presently accelerating."

The authors stated that extinctions of species did not increase during the past 200 years, pointing out, "Past extinctions strongly suggest that climate change is not an important threat to biodiversity. Remarkably, we found here that species-level extinctions related to climate change have not significantly increased over the last approximately 200 years."[46]

So we can put our fears about a mass extinction to rest.

"Climate change causes ocean acidification and destroys corals"

The ocean's ability to absorb carbon dioxide is limited by various factors, including temperature, salinity and nutrient availability. Much of the excess CO_2 is absorbed by terrestrial ecosystems, particularly forests and soils. The ocean has a significant buffering capacity against changes in pH, mainly due to the presence of carbonate and bicarbonate ions. These ions act as a natural buffer. The pH, "potential hydrogen" of a neutral substance is 7.0. Any number lower than 7.0 is an acid and above 7.0 is a base. The pH of the ocean is about 8.1, making it slightly basic, or alkaline. So the term "ocean acidification" is another misnomer.

The Great Barrier Reef, near the east coast of Australia, experienced bleaching events in 2016, 2017 and 2020. This is not an unusual occurrence. It was caused partly by the runoff of chemicals into the ocean from agriculture and sediment from coastal development. A few cyclones occurring during that time period also damaged the reef. However, it has made a wonderful recovery since then and is looking beautiful.[47]

When carbon dioxide enters the ocean, it reacts with the water, forming either acidic carbonate (CO_3) ions, 10%, or basic bicarbonate (HCO_3) ions, 90%. The substance actually used by the corals **is** the bicarbonate ion. This is a blessing.

The Great Barrier Reef qldtravel.com.au

Please don't take our word for this. Check out these facts for yourself.

---

1. https://www.worldwildlife.org/stories/polar-bear-population-decline-a-wake-up-call-for-climate-change-action
2. Amrutha Gayathri, "Polar Bear Population Higher than in 20th Century: Is Something Fishy about Extinction Fears? International Business Times, August 2, 2011
3. www.arcticwwf.org/wildlife/polar-bear/polar-bear-population/
4. IUCN/SSC Polar Bear Specialist Group, http://pbsg.npolar.no/en/index.html
5. Posted on 3/23/23
6. Ibid, p. 16
7. R.L. Holz, "How Many Scientists Does it Take to Write a Paper? Apparently Thousands," The Wall Street Journal, August 10, 2015
8. dso, op. cit., p. 18
9. Tol (2013), Montford (2013), Friends of Science (2014) and Legates et al (2015)
10. Idso, op. cit.
11. http://www.petitionproject.org
12. Hal Lewis, "Hal Lewis Resignation Letter From the American Physical Society," https:// www.heartland.org, Dec. 8, 2010

13. Hal Lewis, "Hal Lewis Resignation Letter From the American Physical Society," https:// www.heartland.org, Dec. 8, 2010

14. https://financialpost.com/opinion/open-climate-letter-to-un-secretary-general-current-scientific-knowledge-does-not-substantiate-ban-ki-moon-asssertions-on-weather-and-climate-say-125-scientists

15. https://www.investors.com/politics/editorials/active-volcanoes-melting-west-antarctic-ice/

16. "NASA – The Mystery of Antarctic Cooling," YouTube video, Jan. 25, 2016

17. https://www.nasa.gov/feature/goddard/nasa-study-mass-gains-of-antarctic-ice-sheet-greater-than-losses

18. Laura Paddison, "Antarctic sea ice hit record lows again. Scientists wonder if it's 'the beginning of the end'", CNN.com, Feb. 2, 2023

19. Tony Heller, op. cit.

20. Professor Emeritus Don Easterbrook's eye-opening talk to the Washington State Senate committee on Climate Change, March 26, 2013

21. Ibid.

22. "Antarctic Ice Shelf Fast Collapsing," *Los Angeles Times*, March 20, 2008

23. "Ice Shelf Hangs by a Thread, *Sydney Herald*, March 27, 2008.

24. "Bye bye, Antarctica?" *Salon* online news, March 26, 2008.

25. The Daily Mail, July 2016

26. Dr. Tim Ball, Human Caused Global Warming: The Biggest Deception in History, 2016, Tellwell Talent, p. 8

27. Seth Borenstein, " 'Arctic is screaming,' says scientists seeing new data", TheHeraldBulletin.com, Dec. 11, 2007

28. Chaffin Mitchell, "NASA study reveals astonishing growth of one Earth's previously fastest-shrinking glaciers," AccuWeather.com

29. Tony Heller, "The 'Ice-Free Arctic' Big Lie", https://real climatescience.com/2016/10/the-ice-free-arctic-big-lie/, Oct. 14, 2016

30. Suwanarat, Sompote, "Recent Cooling Trend in the Arctic and its Possible Causes, *Climate Dynamics*, 2015

31. https://www.noaa.gov/explainers/tracking-sea-level-rise-and-fall

32. Chris Bentley, "Facing rising sea levels and greater insurance risk, Southern Florida braces for relocations, new flood design standards, and more," https://archpaper.com/2017/04/south-florida-rising-sea-levels, April 12, 2017

33. Dr. Tim Ball, "Sea Level Rise; A Major Non-Existent Threat Exploited by Alarmists and Politicians", WattsUpWithThat.com, July 21, 2018

34. "Study Reveals 25 Major US Cities Are Slowly Sinking", *Sweden Herald*, May 11, 2025

35. Nils Axel Mörner, http://scienceandpublicpolicy.org/images/stories/ papers/reprint/sea_level_ not_rising.pdf

36. Mörner, 2007ac, 2010b; Murphy, 2007

37. Newman, op. cit.

38. Kim Bryan, “10 Worst Hurricanes in United States History,” https://owlcation.com/stem/Worst-Hurricanes-United-States-History, Jan. 6, 2017

39. Neil L. Frank, "Did Global Warming Make Hurricane Ian Intensify Faster than Normal?", wattsupwiththat.com, Oct. 6, 2022

40. https://eng-resources.uncc.edu/failurecasestudies/dam-failure-cases/new-orleans-hurricane-katrina-levee-failures/ (ASCE Review Panel 2007, p. v)

41. Malcolm J. Bowman, “Superstorm Sandy – How did it happen and are we prepared for the future?”https://seagrant.sunysb.edu/media/sandy 12/UUPInsight-Sandy020113.pdf

42. Dr. Roger Pielke, Jr., Congressional Testimony, July 18, 2013

43. Giglio et al, 2013

44. Gerardo Ceballas et al, "Accelerated Modern Human-Induced Losses: Entering the Sixth Mass Extinction," *Science Advances,* June 19, 2015.

45. Kristen E. Saban and John J. Wiens, "Unpacking the Extinction Crisis: rates, patterns and causes of recent extinctions in plants and animals, *Proceedings of the Royal Society B.,* October, 2025.

46. Ibid.

47. Dr. Peter Ridd, "Latest 2025 Great Barrier Reef Coral Data: How Bad is it Really?" Reef Rebels YouTube Channel, Aug. 9, 2025

Chapter 7

Cashing In On Climate Change

> "Achieving net zero emissions will require a whole economy transition — every company, every bank, every insurer and investor will have to adjust their business models. . .This could turn an existential risk into the greatest commercial opportunity of our time." – Mark Carney, UN Special Envoy for Climate Action and Finance in 2020 & currently Prime Minister of Canada

Many people are aware that a tremendous amount of resources are being used up all over the world in the effort to mitigate climate change. Not just billions but **trillions** of dollars have been spent on this. The amount increases substantially every decade. During the 1980s and 1990s $50-$100 billion was spent. From 2000 to 2010, the amount spent was $500–$600 billion. From 2010 to 2020 this amount had jumped to $2.3 trillion (million million). From 2020 to 2025 the estimate was $8-$10 trillion US.[1]

Stephen Moore of the Heritage Foundation calls the machine pushing this scam the Climate Change Industrial Complex. Many people are getting extremely rich off of this scam. Because that money is coming out of our pockets, we need to investigate how this has happened. Dr. Ed Berry warned that $1 trillion is wasted every 100 days in the US on climate change mitigation, for a staggering total of $3.65 trillion each year. The United States government took in $4.9 trillion in 2024.

There is a powerful financial incentive for scientists to conclude that an impending climate crisis is in the works. Without fires a fireman wouldn't have a job. Without an impending climate crisis, most climate scientists wouldn't be receiving large grants from governments and other institutions. They wouldn't have jobs.

It's interesting that there are now thousands of climate scientists. Fifty years ago there were very few. At that time the climate was studied mainly by physicists and meteorologists. Fifty years ago universities didn't offer majors in climate science. Now someone can get a master's degree in climate communication.

Science Direct analyzed a dataset of competitive research grants between the years 1950 and 2021. There were 4.3 million awards totaling $1.3 trillion.

As Stephen Moore pointed out, for the huge amount of money that's already been spent on climate change, "We could feed everyone on the planet, end malaria and provide clean water and reliable electricity to every remote village in Africa.

And we would probably have enough money left to find a cure for cancer and Alzheimer's."

Now the climate change "experts" are telling us that between now and 2030, another 5-year period, almost five times that amount, or $7.4 trillion will need to be spent **every year**. They claim that this total of $37 trillion will be required to prevent the temperature increase from exceeding their magic number: 1.5° C.

Cap and Trade

"Cap and Trade", or emissions trading, is a government-sponsored regulatory program designed to limit, or cap, a company's emissions of a gas. A government sets a limit, or cap, on the amount of emissions a company is allowed to produce without paying a penalty. This cap is then gradually reduced as the years go by. Each year the company is allowed to emit less and less gas without a penalty.

Of course, bureaucrats decide the amount of these caps, giving them tremendous power over the companies emitting CO_2. Making such decisions requires a great deal of expertise. Yet there is no universal training program in place for regulatory agencies to translate laws into such operational limits.

Each cap is divided into units for which the government agency issues permits, called carbon credits, or allowances. Each carbon credit is a certificate that allows the company to emit one metric ton (1000 kg) of carbon dioxide gas.

The company must obtain enough of these carbon credits, or allowances, to cover the total amount of emissions they produce, either from the government or by trading with other firms. Sometimes these credits are even auctioned off. This is what the "trade" part in "cap and trade" refers to. If a company has been able to lower its emissions particularly well, it can sell its spare allowances, called "banking allowances" to another company or save them for future needs.

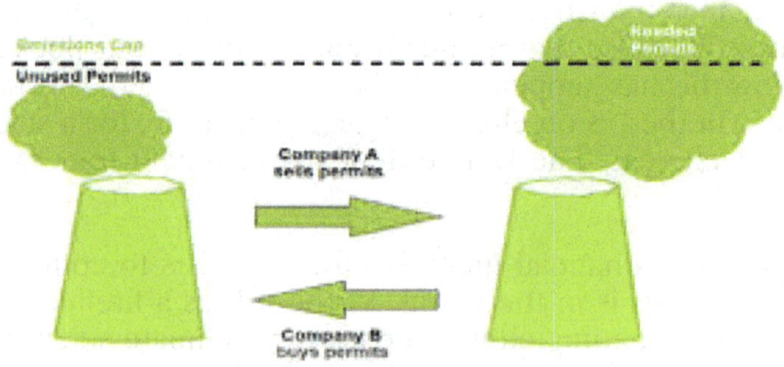

The Elements of Cap and Trade

So if your company emits less gas than the credits you were allowed, you can keep your unused credits or sell them to a company that needs them.

The first cap and trade program involved acid rain. During the 1970s and 80s there was a big media-fueled acid rain scare. The media told us that sulfur dioxide and nitrous oxide gases emitted by utility companies in the Midwest US were damaging forests and lakes in the eastern US, since air masses in the United States move from west to east. NO_2 and SO_2 react with water vapor in the atmosphere to form nitric acid and sulfuric acid. They are real pollutants. Carbon dioxide is not a pollutant.

In 1980 the United States government set up a 10-year Acid Precipitation Assessment program costing half a billion dollars to investigate this problem. Alarm over acid rain that was spread by the media allowed governments to pass regulations governing the sale of sulfur dioxide permits.

Realizing how environmental concerns could generate big bucks, Senators Al Gore, John Heinz and Timothy Wirth introduced a bill known as "Project 88" in June of 1988. This legislation was geared to create a whole new market for selling sulfur dioxide permits. Project 88 morphed into the Clean Air Act, passed in 1990. Although the media discussed this nonstop, these acid rain fears proved to be largely unjustified. Just two species of trees found at a high altitude were damaged. The acidity of the water in lakes was later attributed to natural causes. The scientists who studied this problem found no easy answers.

1988 was the same year that the United Nations Intergovernmental Panel on Climate Change (IPCC), were set up by Maurice Strong, the real mastermind of the global warming scam. Agenda 21 was also set up at that time. In the United States a congressional hearing on global warming set up by Al Gore and Tim Wirth took place on June 23, 1988.

The Enron company, a corrupt energy corporation that went bankrupt in 2001, was the major trader in the $20-billion per year cap and trade market for sulfur dioxide. Because Enron derived so much benefit from these trades, it was looking around for other commodities that could be traded in that way. Due to the concern about carbon dioxide being generated by the 1988 congressional hearing and nonstop media blitz, adding CO_2 to this list seemed like a dream opportunity to the Enron executives. Enron operated a vast network of natural gas pipelines. Because natural gas emits less CO_2 than coal plants, they could make billions of dollars from the coal companies in a cap and trade setup.

The Enron executives worked with Al Gore, Tim Wirth, James Hansen — who had been hired as a consultant — and John Kerry, who later became Biden's climate czar. They lobbied Congress and the Clinton administration to produce propaganda demonizing carbon dioxide so that it could be declared a pollutant. CO_2 was declared a pollutant by the Supreme Court in 2007 and by the Environmental Protection Agency in 2009.

The cap and trade market for carbon dioxide gas was established in 1997 by the Kyoto Protocol. Passed on Dec. 11, 1997, the Kyoto Protocol required industrialized nations to cut their "carbon" emissions by an average of 5% below 1990 levels, which were lower than 1997 emissions. It set binding targets to reduce emissions for 37 industrialized countries and the European Union, called Annex B countries.

Under the Kyoto Protocol the Clean Development Mechanism (CDM) was also set up to allow developed countries to invest in emission-reduction projects in developing countries. For this they could earn "certified emission reduction credits" (CERs) that would meet their own greenhouse gas reduction targets. However, the Kyoto Protocol lacked strong mechanisms for motivating and ensuring compliance. No nation has actually been able to fully comply with it.

The four largest cap and trade systems currently in operation are:

1. The European Union Emissions Trading Scheme, (EU ETS) established in 2005, with about 11,000 participants.

2. The National Emissions Trading Scheme, in China with about 2,200 participants.

3. The Korea Emissions Trading Scheme, (E-ETS) the first mandatory east Asia trading system, with 684 participants, and

4. California's Cap and Trade Program, started in 2013, with about 450 power plants and fuel distribution organizations.

Carbon Offsets

In 1989 the World Resources Institute (WRI) launched the first carbon offset program. This program involved planting trees. They figured that the trees would absorb the carbon dioxide that the polluters were responsible for emitting. However, when they are first planted, trees sequester hardly any carbon dioxide at all. They don't manage to absorb much CO_2 until they are about twenty years old. Mature and old-growth trees are really the powerhouses of carbon dioxide storage.

The idea of carbon offsets and trading had been toyed with even before 1989. Voluntary and mandatory plans were devised to allow companies to purchase carbon credits from projects designed to offset (balance) CO_2 that they couldn't avoid emitting. Carbon offsetting is a huge industry employing thousands of people to manage these programs.

There are several companies managing carbon offsets. Verra, Climate Partner, Rye Strategy, United Energy Trading, and Southstar Energy Services are a few of these. The largest company running this system is Verra, which has issued offsets for companies such as Disney, Shell, Gucci, and other big corporations. Verra has established standards for carbon offsets and has administered three-fourths of them. Using Verra's algorithms, other organizations working on these projects estimate how much carbon dioxide a certain project will save. They figure out how many trees will absorb a certain amount of carbon dioxide and promise not to cut down those trees for a set amount of time, perhaps 50 years.

The Guardian reported a nine-month investigation — using satellite images — which found that over 90% of rain forest offsets issued by Verra were for saving stands of trees that didn't actually exist. These are called "phantom credits". Two other studies concluded that over 94% of these offsets didn't benefit the climate at all. *The Guardian* reported, "Emissions trading was supposed to save the planet. But fraudsters quickly learned how to rip the system off, making themselves spectacularly rich. Then some of the major players started turning on each other."[2]

Another article about this scam appearing on the Greenpeace website described the COP-29 conference in Baku. "The last few years have seen scandal after scandal in the schemes of false solutions, from ineffective projects to exploitations against indigenous peoples."[3]

As might be expected, some of these offsetting projects have seriously violated human rights. *The Guardian* viewed videotape evidence that, as part of one of these projects, people in Peru were evicted from their homes which were then cut down with chainsaws and ropes by park guards and police.

In these reforestation projects local governments have been enticed to impose hefty fines of thousands of dollars on their citizens for cutting trees down. This practice, called neocolonialism, has had a negative impact on local people's daily lives. Farmers in Brazil lamented that they don't have enough lumber to build their homes. "People want us to starve to reduce carbon emissions," they cried.

Georgetown University professor and philosopher Olufemi Taiwo believes that richer countries are dominating less powerful countries and people through these initiatives meant to slow the pace of climate change. He calls this new form of neocolonialism "climate colonialism". The Global South is a group of nations with a relatively low level of economic and industrial development. They are typically located to the south of the more industrialized nations. Rich countries are investing in the Global South and dictating how to manage land and forests to people who have lived there for thousands of years in many cases.

At the Glasgow COP-26 conference in 2020 indigenous attendees declared that the climate solutions under discussion were not only false, but that they extended colonialism using land grabs and displacement of native people in the Global South. Luis Arce, President of Bolivia, pointed out important concerns about how the UN uses carbon offsetting to impose rules on marginalized countries by leaders of powerful, industrialized countries. He calls this practice the "new carbon colonialism".

A parallel conference, called the "People's Summit for Climate Justice" was held near the convention center where COP-26 was being held to call attention to these injustices.

Reducing Emissions from Deforestation and Forest Degradation (REDD+) is another United Nations-backed program designed to mitigate climate change by preventing forest destruction. But a 2022 University of Cambridge study revealed that in these projects Verra had overstated the threat to forests by an average of 400%. In the booming trade of carbon offsets sellers are flooding the market with worthless products. "We can see that saving vast swaths of forests from the chainsaw to balance emissions are overblown," declared Professor Andreas Kontoleon, from Cambridge's Department of Land Economy. This is a big business from which undeserving people are getting filthy rich.

Subsidies for "Renewables"

A huge amount of taxpayer money earmarked for climate change disappears into subsidies, which are gifts or grants of money to an industry to encourage its existence. In doing this governments influence purchases, propping up industries like ethanol, solar and wind power. This is how politicians can pick winners, lobbyists who support their campaigns. Without subsidies these industries couldn't compete in the free market. This gives governments tremendous power. Billionaire Warren Buffet described the situation in his honest disclosure, "We get a tax credit if we build a lot of wind farms. That's the only reason to build them. They don't make sense without the tax credit."

Without subsidies and mandates intermittent wind and solar could never compete with reliable energy like coal, natural gas and nuclear energy, which still provide about 80% of all the electricity generated in the United States. Wind turbines and photovoltaic panels are not renewable, because they are made using "fossil fuels". They last only about 30% as long as coal and natural gas generating plants. Fiberglass wind turbine blades cannot be recycled. They must be cut up and buried in the ground.

Even with subsidies some of these wind and solar projects haven't been able to survive financially. A case in point is Ivanpah, located in the Mohave Desert. Its thousands of mirrors reflect sunlight onto a tower filled with a fluid that heats up and generates steam to run a turbine. This plant never worked as well as had been advertised. Ivanpah quickly became known as a bird-killing eyesore. About

5,000 birds have been killed by heat reflected from its mirrors. These poor birds actually catch on fire. Also, this project had taken over a tortoise habitat. It is scheduled to close in 2026.

Ivanpah Solar Plant   Credit: Inhabitat.com

In 2011 the US Department of Energy (DOE) gave Ivanpah a $1.6 billion loan guarantee. Ivanpah's owners were exempted from the standard credit subsidy cost and were granted several loan extensions. These loans have still not been fully repaid. Then in 2014 Ivanpah received a $539 million grant from the Treasury Department. Despite this massive financial lift, Ivanpah failed to meet its energy output goals. Benjamin Zycher of the American Enterprise Institute, notes that when "fully operational it was supposed to generate about one million megawatt-hours of power annually; from 2015 to 2023 actual annual output has averaged about 700,000 MWh."

Treasury Department figures show that subsidies for wind and solar dwarf all other energy-related provisions in the tax code, costing about $31.4 billion in 2024 alone. Future expenditures to taxpayers from 2025 to 2034 are estimated to be $421 billion — almost half a trillion dollars. In the last ten years alone subsidies for wind and solar increased by 21 times. The cost of the Investment Tax Credit (ITC) and Production Tax Credit (PTC), the primary drivers of wind and solar energy deployment, are the most expensive energy-related provisions in the federal tax code.

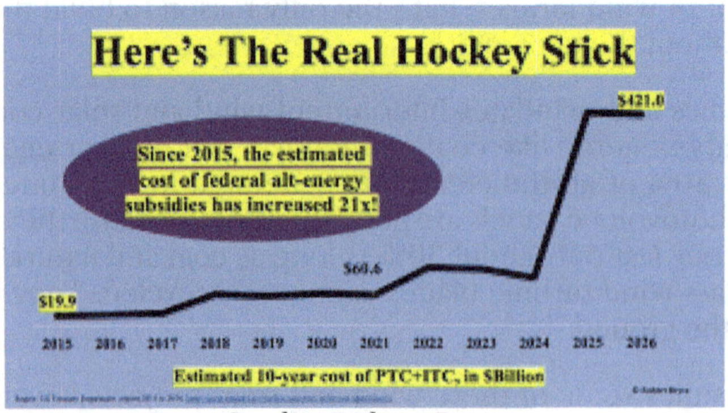

Credit: Robert Bryce

The Inflation Reduction Act

On August 7, 2022, the US Congress passed the "Inflation Reduction Act" (IRA) on a 51-50 vote, with all of the Democrats voting in favor of this bill, all Republicans voting against it, and Vice President Kamala Harris breaking the tie. Of course this bill accomplishes the opposite of its name — another lie.

Research by the Cato Institute indicates that the IRA's total "green energy" subsidies will likely cost between $2 trillion-4.7 trillion by 2050, about three times the cost originally projected by Congress. Although some congressional representatives want to repeal the IRA, the Institute for Energy Research doesn't believe these subsidies will end anytime soon.

Added to that expense is the money spent by 29 states that have mandates forcing public utilities to buy expensive "green" energy requiring 400-1,000 times as much land as conventional power plants for every unit of power generated. This doesn't include the tax credits for electric vehicles, projected to cost US taxpayers $105.7 billion between 2025 and 2034. Only the rich can afford battery-powered vehicles. Those who don't have the money to buy them are helping to support the rich with their tax dollars. The owners of EVs are exempted from the federal road tax that drivers of cars powered by internal combustion engines are required to pay. This is ironic, because EVs are heavier than internal combustion vehicles and thus cause more damage to roads.

A case in point about the devastating effect such policies can have on a nation's economy is what has happened to Germany. The new US energy secretary, Chris Wright, noted, "Germany spent half a trillion dollars, made their energy two to three times more expensive, and they produce 20% less electricity today than they did 15 years ago. The Biden regime was following the same path. We're not going to go down that road. We want affordable, reliable, secure energy and a reindustrialization of America, not deindustrialization of America."

Carbon Taxes

Governments were quick to realize that the climate change "problem" provided another great opportunity to rake in money. Taxes on energy affect the world's entire economy. They are not just a tax on CO_2 emitters. All energy taxes on these companies are passed on to consumers as inflation and increased energy costs: higher prices at the gas pump and higher electricity prices to heat homes and businesses. Energy, whether fuel or electricity, is the foundation of virtually every step of the supply chain.

The transportation of food from farmers' fields to factories where it is processed and then to grocery stores is also powered by petroleum. These taxes on energy producers raise prices at every step of the supply chain. This has a compounding effect. The final cost of a product you buy at a shopping mall or grocery store has been increased multiple times by these energy taxes. It was estimated that energy taxes raise the cost of each part of a supply chain 64% faster than the rest of the economy.

The Institute for Energy Research reported that in Canada the Trudeau government instituted a carbon tax in 2019 starting at $20 per metric ton of carbon dioxide emitted by a power company or airline. By April of 2024 that tax had increased to $80 per metric ton — four times as much. By the year 2030 this tax is scheduled to be $170 per metric ton, more than eight times as much. Residential electricity customers found that this carbon tax actually cost them more than the fuel they were buying to heat their homes.

In Ontario, where low temperatures are often -29°C (-20° F) in the winter, one woman's gas bill included a carbon tax of $134. In Peterborough, Ontario, a restaurant was billed a carbon tax of $1,044 in addition to their regular heating bill. A man in Alberta was charged a carbon tax of $136 for one month in addition to the cost of natural gas. He also had to pay a carbon tax of $0.17 per liter of gasoline for his truck. Because he uses 200 liters of gasoline per week, this adds $34 extra per week, amounting to $1,768 per year for this tax. These exorbitant and unfair taxes have forced some families in Canada to take up residence in campers to save money on their heating bills.

In 2012 a $25 per ton carbon tax was introduced in Australia. This 19-year plan would have cost $162 billion. The goal was a modest 5% reduction in carbon dioxide emissions. This measure caused such extreme economic deprivation that the number of bankruptcies skyrocketed. The tax had to be discontinued. Because Australia's CO_2 emissions amount to only 1.2% of the world's total carbon dioxide emissions, this reduction would have made a difference in the earth's average temperature of just 0.00005 (5/100,000) degrees Celsius after ten years. And that's only if the IPCC figures are correct.

If we extend those figures out to worldwide proportions, we find that the cost of reducing the earth's temperature by only one degree Celsius would be $3.2 quadrillion ($3,200 trillion). This is assuming that the IPCC figures are correct. This cost, spread out equally among the eight billion people on Earth, would amount to $400,000 US for every man, woman and child.

Net Zero Alliances

A "net zero alliance" is any global group of financial institutions, asset owners or other stakeholders committed to achieving a net-zero economy by 2050. The Net Zero Banking Alliance (NZBA) is a subset of GFANZ, the Glasgow Financial Alliance for Net Zero, the world's richest climate alarmist organization. Headed by Michael Bloomberg, GFANZ controls collective assets-under-management of about $130 trillion of other people's money.

The Forum for Insurance Transition to Net Zero (FIT) is a UN-led multi-stakeholder forum to support the "acceleration and scaling up of voluntary climate action by the insurance industry and key stakeholders." [4]This may be a ploy to change the wording of their strategy to protect it from antitrust lawsuits. There's been a trend towards higher insurance rates, especially in the solar industry, where property insurance rates have risen by up to 45%. Insurers have become more selective about which solar manufacturers they will insure, citing manufacturing defect claims as a concern.

Climate Action 100+ is the world's largest and most aggressive investor initiative on climate change. Its 700 investors are intent on insuring that 166 of the world's biggest corporate "greenhouse gas" (GHG) emitters take the necessary actions to align their business strategies with the goals of the Paris Agreement.

Since December of 2024 something really amazing has been happening. Bank of America, Citigroup, Goldman Sachs, JP Morgan, Morgan Stanley, and Wells Fargo left the Net Zero Banking Alliance in an apparent reaction to (environmental, social and governance) policies. Perhaps they can see that there is not much future for this industry. Perhaps it's becoming apparent to many that none of the climate predictions and tipping points made during the last four decades has come to pass. And alternative energy is not living up to its claims.

BlackRock is a global investment management company, the world's largest asset manager, with $11.6 trillion in assets under management as of March 31, 2025. In November of 2022 eleven state attorneys general in the United States filed suit against BlackRock and StateStreet for violating antitrust laws. These asset managers, working together, acquired enough stock in every major coal company in the United States to dictate their policies. Using Climate Action 100+, they pressured these coal companies to fall in line with "green climate" goals, to cut their coal production by 50% by the year 2030. This deliberate reduction in the supply of coal increased energy prices in the United States, while at the same time producing dramatic profits for these asset managers.

These companies are being investigated under the lens of the Sherman Act, an antitrust law passed in 1890. This law was designed to prevent companies from engaging in practices that stifle competition, such as price fixing or limiting output. It prohibits contracts or conspiracies that restrain interstate or foreign trade and attempts to monopolize.

Now BlackRock has also left the Net Zero Coalition of Corporations that had pledged to reach Net Zero by 2050. BlackRock's manager explained that membership in the group "caused confusion regarding BlackRock's practices and subjected us to legal inquiries from various public officials."

These collaborations restrict competition by influencing investment decisions, limiting the supply of "fossil fuels" and otherwise interfering with market forces. These companies have also colluded to exert collective pressure on businesses to reduce their emissions by withholding or threatening to withhold capital from them.

In May of 2024 Climate Action 100+ lost three of the world's six largest asset management funds: BlackRock, JP Morgan Chase and StateStreet. Some asset managers were concerned that this cartel was pressuring other companies too aggressively. BlackRock, StateStreet and JP Morgan Chase together control about $18 trillion in funds.

The UN Environment Program claims that its 450 members represent more than 100 trillion dollars that are pledged to promote climate mitigation. This program appears to have been set up by the World Economic Forum (WEF). We know that the WEF has essentially joined with the UN. Is it any wonder that state attorneys general believe that antitrust laws have been violated?

Unfortunately it doesn't seem that Canada's new prime minister, Mark Carney, who claimed to be an outsider before the recent election, will be of much help. In 2020 Carney was the UN Special Envoy for Climate Action and Finance. He recently commented, "We've retooled the financial system. We've changed the plumbing, if you will. The private sector needs to step up and say they'll help finance the 100-trillion-dollar transition that needs to happen over the next three decades. We need organizations like GFANZ that are relentlessly, ruthlessly, absolutely focused on Net Zero."

Climate Lawfare

"Lawfare", a type of warfare in which the law is used as a weapon against your opponents to damage or delegitimize them, wasting their time and money and gaining public relations points has been increasingly used by governments and climate activists. If activists can't get their way, they take their issue to a sympathetic court. This practice is now threatening the prosperity and technological advancement that has come from our use of hydrocarbons. Again, the financial suffering is inevitably passed on to the consumer. Yet the perpetrators of such litigation — mainly state governments and climate activist groups — offer us no sustainable alternative to "fossil fuels".

Because climate change is a global issue crossing international borders, any such legal action should be the domain of the federal government, not states.[5]

Just since 2015 the number of climate litigation cases around the world has doubled, with over 1,000 new cases being reported.[6] These cases, while mostly in the United States, are occurring in nearly forty nations. The American Bar Association seems to be encouraging these. "For any lawyer interested in corporate accountability, human rights, the environment or the fate of the world, climate litigation is an area to watch and engage in."

These cases often allege "deceptive conduct" and "misrepresentations of climate science". Oil companies are also accused of "negligence, trespass, failure to warn, design defect, conspiracy and unjust enrichment".[7] In a decade-long case started in 2014 in Louisiana, trial lawyers steered almost $15,000 to the judge overseeing the case. The judge, Michael Clement, then ruled in favor of the plaintiffs and ordered Chevron to pay $744 million. Former attorney general Bill Barr described this as a "kangeroo court". This case is being appealed.

Some of the most notorious cases on the books are:

Comer v. Murphy Oil USA, Inc.: In 2011 homeowners in Katrina-ravaged Mississippi sued energy companies such as Chevron and ExxonMobil, alleging that their emissions contributed to global warming, which in turn caused Hurricane Katrina. The case was dismissed on the grounds that federal common law did not recognize a cause of action for climate change impacts. Unfortunately, however, this case paved the way for future climate change lawfare litigation.

Kivalina v. ExxonMobil Corp et al. (2008–2013): The Alaskan village of Kivalina sued 24 energy companies, alleging that their greenhouse gas emissions contributed to the melting of sea ice, threatening the village's existence. The case was dismissed on the grounds that the court lacked jurisdiction.

Alec L v. Brown (2015): In this case 21 young plaintiffs, represented by Our Children's Trust, sued the state of Washington and fossil fuel companies, arguing that the state's failure to reduce greenhouse gas emissions violated their constitutional rights. The case was dismissed, but it brought attention to the intergenerational aspect of climate change.

City of Baltimore v. P.L.C. et al (2018-present): The city of Baltimore sued 26 fossil fuel companies, alleging that their products contributed to climate change causing harm to the city's infrastructure and economy. The case is ongoing and has been consolidated with similar lawsuits from other cities and counties.

State of Rhode Island v. Chevron et al (2018-present): Rhode Island sued 21 fossil fuel companies, alleging that their products contributed to climate change, causing harm to the state's infrastructure, economy and natural resources. The case is ongoing and has been consolidated with other similar lawsuits.

Amazon Rainforest Communities v. Chevron Corporation (2019-present): Indigenous communities in the Amazon rain forest sued Chevron, alleging that the company's activities in the region contributed to climate change, threatening their way of life and the environment. The case is ongoing and seeks to hold Chevron accountable for its alleged role in climate change.

"Sustainable" Investing

For many years the World Economic Forum (WEF) has advocated for "stakeholder capitalism", pushing governments and large corporations to pass laws mandating stricter climate regulations and huge subsidies for "green" energy, via ESG. Thus Europe has lost its manufacturing base to China. [8] Trillions of dollars have moved from the public to bankers and "renewables" investors, the greatest transfer of wealth from the poor to the rich in history. Despite all of this, "fossil fuel" use increased by one third between 2000 to 2025.

On November 19, 2023, Shane Shifflet of *the Wall Street Journal* noted, "Wall Street rushed to embrace sustainable investing just a few years ago. Now it is quietly closing funds or scrubbing their names after disappointing returns that have investors cashing out billions."The third quarter was the first time more sustainable funds liquidated or removed criteria from their investment practices than were added, according to Morningstar. That is a reversal from not that long ago when companies were rebranding nonfaltering funds to cash in on the billions of dollars flowing into sustainable investment products."

Since 2018 some business executives have been warning that the wind and solar industries would ultimately collapse due to their continued reliance on subsidies. *Venture Capitalist* predicts, "An Enron-style collapse is in store for renewables." Despite these gains for the side of truth and justice, these politicians and climate investors won't go down without a fight. The push for communism won't be so easily relaxed. These misguided and insane policymakers won't admit that they were wrong and that the energy alternatives they've cast upon the world are an economic disaster.

---

1. "Global Climate Mitigation Finance Growth", Grok.com, Jan. 2026
2. Nick Davies, "Abuse and Incompetence in Fight Against Global Warming," *The Guardian*, June 2, 2007.
3. Iqbal Demonik and James Stoppel, "Trading Carbon Offsets Won't Save the Planet, Time for Real Climate Solutions, Greenpeace, Nov. 13, 2024.
4. Louise Istad, "UN Launches Forum for Insurance Transition to Net Zero," Insura nceday.com, April 25, 2024.
5. Steven Bucci, "Local Climate Lawfare is Undermining US Foreign Policy and National Security", *The Daily SIgnal*, Nov. 19, 2025.
6. Benjamin Franta, "Climate litigation riding; Hot spots to watch," American Bar Association, Dec. 22, 2021.
7. Ibid.
8. Michael Shellenberger, "WEF Spread Disinformation About Capitalism to Concentrate Wealth and Raise Prices, Says Deutsche Bank Whistleblower," Climate Depot, Jan. 22, 2026.

Chapter 8

The Deception Machine: Climate Change Propaganda

> "We will know our disinformation program is complete when everything the American public believes is wrong." — William Casey, Former CIA Director

"The impacts of climate change are already being felt in communities across the country."[1] This seemingly innocent statement raises a red flag. We need only examine the documented evidence to realize what a false claim this is. Despite what you may have heard on TV or read in articles, satellite temperature records show that there has been no appreciable warming in the past 24 years.

Statistics indicate that the incidence of serious hurricanes, wildfires, tornadoes, drought, and floods is about the same or even less than before the percentage of the trace gas carbon dioxide in the atmosphere increased. The rate at which the sea level is rising is about the same as it has been for the last 1000 years: seven inches (not twenty feet!) per century. There is no evidence that the ocean is becoming acidic. Ocean water has an average pH of 8.1 and is not acidic but alkaline.

Al Gore, "An Inconvenient Truth"

Hurricane Katrina, a deadly hurricane that made landfall in 2005, was featured heavily in Al Gore's documentary, "An Inconvenient Truth", which aired in 2006, just a year later. This hurricane was still fresh in everyone's memory, giving the film a tremendous impact. However, looking at the entire record, we know that the incidence of major hurricanes has actually decreased in recent years.

In 2007 Gore received the Nobel Peace Prize for this documentary at about the same time that a British judge, Michael Burton, listed nine significant errors in the "one-sided" production and ruled that it could be shown in British schools only if accompanied by teacher guidance explaining these errors to their students. Among other serious errors, Gore claimed that melting in West Antarctica or Greenland would cause a twenty-foot sea level increase in the near future.[2]

In this documentary Gore cited the fact that in 1895 Svante Arrhenius had published a paper naming carbon dioxide as a greenhouse gas and predicting that if its percentage were to double, the earth's temperature would rise by 6 degrees Celsius. However, Gore failed to mention that in 1906 Arrhenius re-examined his figures, discovering that he had made an error. Arrhenius then corrected his prediction: that if CO_2 doubled, the temperature would rise by two degrees, not six. Surely Gore — or whoever wrote that documentary — was aware of that fact. But six degrees sounds more threatening than two degrees. [3]Of course, purposely withholding information is a form of lying.

Nowhere is this outrageous propaganda more evident than in our schools. Just as the Nazis brainwashed the German children, the global warming story, which hijacked environmentalism, has unfairly taken advantage of captive audiences in our schools to influence their young minds.

On March 15, 2019, school children all over the world took the day off from school to participate in a demonstration to "save the world from climate change." The headmasters of schools in Britain suspended classes even though their students were supposedly "on strike". It was obvious that this affair, a PR stunt arranged to garner more media attention, was organized by the adults.

This manipulation of young children's minds takes advantage of the fact that children are impressionable and lack the education and experience to discern truth from fiction. The strategy of the organizers is akin to that used by the Hitler Youth leaders in Nazi Germany.

A nightmarish example of this type of manipulation is David Suzuki's statement: "The North Pole, once a wintry wonderland, is no longer safe for Santa's Workshop. Climate change is melting the snow and ice, and the rising water is getting too close for comfort. Santa must relocate — fast — to make sure that all the nice boys and girls still have a Happy Holiday."[4] Dr. Patrick Moore calls this disgraceful exploitation of children to advance one's agenda "child abuse". He recently told the climate alarmists in a tweet:

> "You don't seem to get it, that this is a fascist tactic. You are using innocent children for your own ends. And this is child abuse of a psychological nature. It's normally called brainwashing, and you are guilty of participating in it. Mao would be proud. . .You are clearly projecting your own psychosis on the children. Self-loathing is a heavy burden. Why don't you express it through adults rather than abusing innocents?"[5]

Ceaseless propaganda that blames climate change for every adverse weather event is taking a huge toll on children and even some adults. They've given it a name: "climate anxiety". This condition has also been called "eco-guilt", "eco-anxiety" and "eco-grief". This continual assault on the mind has even caused some children to develop post traumatic stress disorder (PTSD). A poll taken in 2022 revealed that 45% of the population are very concerned about bringing children into the world; some have even made the decision not to do so. This has definitely influenced the replacement rate in industrialized nations. It's also helping to destroy the family, which should make the elites calling for population reduction very happy.

A Los Angeles Times headline stated, "It's almost shameful to want to have children."[6] A headline in the New York Times reads, "Climate Change Can Cause Bridges to Fall Apart Like Tinkertoys."

Climate propaganda is a worldwide phenomenon, and certainly this is not an accident. A survey of 10,000 young people in several nations around the world disclosed that 75% of them expressed fear of the future. These young people reported feelings of sadness, anger, helplessness and guilt. Over 45% revealed that climate anxiety was affecting their daily lives. What makes this sad situation so atrocious is that the AGW hypothesis has never been proven.[7]

Propaganda bypasses logic by using faulty reasoning – logical fallacies – and emotional appeal. We have a great responsibility to teach children, as well as adults, critical thinking skills to help them recognize the signs that they are being exploited.

Some logical fallacies (errors in logic) constantly used in man-made global warming propaganda are:

1. “ad hominem” (against the man) attacks, which either directly call names, such as “climate denier” or indirectly disparage those who disagree by assaulting their credentials or character – eg. “on the fringe”.

2. “straw man” - distort your opposition’s argument and then knock it down. A “climate change denier” is someone who denies that the climate is changing. No skeptic ever said that.

3. “band wagon” – states that there is a consensus of the majority promoting the agenda, and that anyone who disagrees doesn’t belong. This works because of the basic human need to be on the “winning team”.

4. “appeal to authority” - We are constantly told to “trust the experts” without evidence. The telltale sign is the lack of information. We are never told that there are other experts who disagree with the “experts”. We are starting to learn from the nonsense that many (not all) people with doctoral degrees are spouting — such as the ability to change one's gender ("trans") — that they lack common sense and the ability to think. Anyone who has passed Biology 101 knows that in every one of the trillions of human body cells (except for red blood cells), each male cell contains an XY chromosome and each female cell contains a XX chromosome. Therefore, despite the administration of hormones and multiple operations changing sex organs, it is utterly impossible for anyone to change to the opposite gender.

5. “moving the goalpost” - When nature failed to cooperate with the global warming models and Earth stopped warming after 1998, the global warming alarmists simply changed their mantra from “global warming” to “climate change” and even

"climate disruption" to falsely imply that this "crisis" was causing storms, droughts, etc. Another way they've moved the goalpost is to keep advancing their "tipping points" to the future after their current tipping point passes without incident.

6. "appeal to sympathy" - blames skeptics for the plight of the polar bear and the victims of storms, droughts, floods and forest fires, which, of course, are no worse than they've ever been.

7. "guilt by association" These experts in psychology have tried to create a **false association** of climate skeptics with folks in groups like the Flat Earth Society, who believe the earth is flat. Of course, no climate skeptic believes such a moronic idea, but the constant repetition of such assertions sticks in the public consciousness after a while.

Other techniques used in this insidious propaganda are:

1. Weather events that are normal are presented as abnormal.[8] The climate events we've experienced in the last century are all within the parameters of natural climate change. The statistical record plainly shows that despite what the alarmists say, there has been **no** increase in droughts, floods, heat waves, wildfires, hurricanes or tornadoes. Some of these have actually decreased.

One example is this headline: "The 2017 Hurricane Season Really Is More Intense Than Normal." The key phrase here is "more intense than normal".

2. Definitions are changed. This is a logical fallacy known as equivocation. For example, in 2004 climate alarmists, realizing that the warming had stopped, changed their mantra "global warming" to "climate change".

This stroke of genius opened up a nearly infinite range of possible things that could go wrong, making the "problem" appear more threatening. Now you see headlines claiming that it is either the warmest, coldest, wettest, driest, calmest, or windiest month or year on record.

3. They play on our emotions. - fear, guilt and pity - rather than reason and logic. They tell us that we should feel guilty for what we are doing to our children and grandchildren. We should take pity on the poor polar bears.

4. They take advantage of the public's lack of science knowledge. Climate is a complicated subject beset with uncertainty, and climate change science is in its infancy. Even some scientists who specialize in other disciplines, such as biology, may lack the confidence needed to do their own research on the climate and make an informed decision. So they, too, trust the "experts". This enables the climate alarmists to change terms such as "polar vortex" to suit their own goals and to make unscientific statements, such as, "The science is settled." Real science is never settled.

5. They keep ramping up the threat to make it appear more and more ominous.

Here are some of the latest threat predictions and other disinformation being propagated by the media:

"Global sea levels are projected to rise between 2 and 7 feet by 2050, and possibly more. By the end of the century, sea levels could rise as much as 20 feet. In this scenario, the United States alone would lose nearly 50,000 square miles, which today are home to 23.4 million people."[9]

> "In 2020, scientists discovered a new source of greenhouse gas emissions off the coast of East Siberia, as warmer waters cause the seafloor to release ice-encased methane, a potent source of global warming. According to the U.S. Geological Survey, the so-called Arctic hydrate destabilization is one of the most serious scenarios for abrupt climate change that has implications far beyond the melting ice of the polar north."[10]

They don't tell you that the amount of methane in the atmosphere is only **2 parts per million**, with an almost imperceptible influence on the climate, making this argument preposterous.

A good example of this type of propaganda is the following editorial by Mark Hertsgaard: "Climate Denialism Is Literally Killing Us. The victims of Hurricane Harvey have a murderer—and it's not the storm."

"The horrors hurled at Houston. . . in late August were heartbreaking — but also infuriating. How many times must we see this disaster movie — titled Hurricane Harvey in 2017, Hurricane Sandy in 2012, and Hurricane Katrina in 2005, along with many lesser known foreign releases — before we intervene and change the ending? And how long before we hold the ultimate authors of such climate catastrophes accountable for the miseries they inflict?

"The tragedy of Harvey starts with the suffering of innocents like Jordyn Grace, the 3-year-old who survived the flood by clinging to the body of her drowned mother, who had prayed with her last breaths. At least 60 people died in Texas because of the storm, over 1 million people were displaced, and who knows how many survived but lost everything? "What makes this so infuriating is that it shouldn't be happening. . .And how long before we hold the ultimate authors of such climate catastrophes accountable for the miseries they inflict?"[11]

The rant goes on and on. To whip people into an emotional frenzy, Hertzgaard blames this storm on "climate denialism", which itself is a misnomer. No one denies that there is a climate or that the climate is changing.

A category 4 storm, Harvey, in 2017, was the country's first major (category-3 or higher) hurricane to make landfall on the US mainland since hurricane Wilma hit Florida in 2005. This ended a twelve-year period, the longest in US history, in which no Category-3 or higher hurricanes hit the coast of the United States. Either Hertsgaard didn't do his homework or he tried to push lies onto the American people in a deliberate attempt to instigate fear and misdirected blame.

This map of the United States appeared in the November, 2013, issue of *National Geographic* magazine. The projected sea level rise used in this map is based on a sea level increase of 80 meters (262 feet) that was predicted in 2000 by the US Geological Survey (USGS). But such a sea level increase could occur **only** if — in addition to the Greenland glaciers and Arctic sea ice — **all** of Antarctica's glaciers melted. If that were to happen, the earth's sea level would rise by ~73 meters (about 240 feet).[12]

The United States if Antarctica Melted

The periods of very slight warming within natural limits that have occurred in the last few hundred years have provided no evidence that anything like this could ever happen for a long time: millions of years. As a matter of fact, we are actually living in an interglacial period: between periods of glaciation in the Pleistocene ice age. So this article in National Geographic magazine is a figment of the imagination and a supreme example of propaganda. The editors of *National Geographic* should be ashamed that they ever published such an article.

Dr. Patrick Moore summarized what has been happening.

> "Fear has been used all through history to gain control of people's minds and wallets and all else, and the 'climate catastrophe' is strictly a fear campaign — well, fear and guilt — you're afraid you're killing your children because you're driving them in your SUV and emitting carbon dioxide into the atmosphere and you feel guilty for doing that. There's no stronger motivation than those two."[13]

In George Orwell's classic, *1984*, a totalitarian government controlled its people with terror and propaganda lies and even by rewriting history. Some of the slogans used by the Party were: "War is peace. Freedom is slavery. Ignorance is strength." The people who had been carefully taught not to question anything went along with this party line. Now we are beset by the frightening realization that the essence of this story is coming true. The public is now being told that the frigid winters we've experienced in the past several years indicate that the earth is warming.

This is a good example of climate change double talk. A recent CNN report claimed that global warming doesn't always mean warming. "The term 'global warming' confuses people, because it triggers thoughts about warmth, and it sort of lends itself to misinterpretation when it also impacts the cold," declared Mike Hulme of the University of Cambridge. The message from this is "Don't trust your senses or your instruments."[14]

When skeptics point to several straight years of severely cold winters, they are reminded that weather isn't climate. "Not only are severe snowstorms possible in a warming climate, they may even be more likely. . .These storms draw much of their intensity from the extreme contrast between cold air over land and warmer, wetter air from over the ocean," asserted a NOAA spokesperson. So we are being told to suspend scientific observation.[15] But a heat wave that is entirely within natural limits is used for an article warning us that climate change is destroying the planet. So whatever happens: whether it's searingly hot or frigidly cold, whether there is too much rain or not enough rain, the relentless propaganda to which we've been subjected leads to the inevitable conclusion that these weather events are all the result of climate change that we have caused.

What's frightening about this propaganda is that so many people believe it without question. Those advancing this barrage of lies to which the world has been subjected for nearly forty years know that, as Adolf Hitler stated, "If you tell a big enough lie and tell it frequently enough, it will be believed." And the purveyors of propaganda are banking on the unfortunate tendency of people to forget past predictions that never materialized.

The corrupt media have been getting away with pumping out these false stories nonstop. Now, after hundreds of cleverly orchestrated false alerts and tipping points, it seems that some people have more or less accepted the concept that climate change is "here" and is "even worse than predicted". For example, we read here: "As climate change becomes more severe than originally projected, estimated adaptation finance needs for developing countries are doubling or tripling every few years."[16]

An unintended blessing of the COVID-19 scam we've just lived through, with its tyrannical lock downs, mask mandates and vaccine mandates, is that now many people are finally awake to the fact that they've been lied to incessantly by governments and the media. Some people wonder if the COVID lockdowns may have been a test to determine how far people will go to obey farfetched commands.

People have been so bombarded day after day by this climate propaganda that it's difficult not to be affected no matter who we are. Since the years of Obama's presidency people in the United States have been unaware of the extent of "news" that is fake. This also isn't an accident. The Smith-Mundt Modernization Act, passed by the US congress in 2012, allowed thousands of hours of misleading radio and television propaganda **funded by the government** to be disseminated on cable and the airways. Now they can get away with totally false stories that every weather event that comes along is caused by human-caused climate change, that the oceans are boiling and that life on Earth will end in 2030.

It doesn't matter that such weather events have been occurring since the beginning of recorded history. But now climate change is blamed for every hurricane, flood, heat wave, cold spell and even for crime, mental illness, alcoholism, plagues of locusts, and crop failures. In some ways this is even worse than what happened with the COVID pandemic.

Shortly before his death author Michael Crichton, famous for *Jurassic Park* and the TV series *ER*, gave a talk in which he compared climate alarmism to a religion for "urban atheists".

> "There's an initial Eden, a paradise, a state of grace and unity with nature, there's a fall from grace into a state of pollution as a result of eating from the tree of knowledge, and as a result of our actions there is a judgment day coming for us all. We are all energy sinners,

> doomed to die, unless we seek salvation, which is now called sustainability."

Unfortunately, the "renewables" being pushed by climate alarmists are anything but sustainable. Even worse, it is the "renewables" which are destroying the environment.

This incorrect pie graph of "greenhouse" gases — just one of the 99.9% found by search engines, has already been discussed in this book:[17]

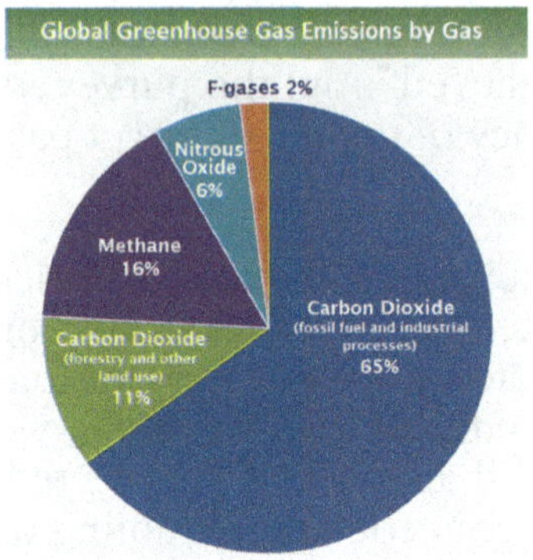

The Main Greenhouse Gas is Missing

Some of the people promoting this lie may not even realize that the ideas they are pushing are wrong, because the lies being presented are everywhere. You can't escape from them. Only after putting in the search term, "CO_2, 0.04% of atmosphere," was the correct pie graph found on line. This is a supreme example of how thoroughly people have been lied to and how completely the students in our schools are being brainwashed.

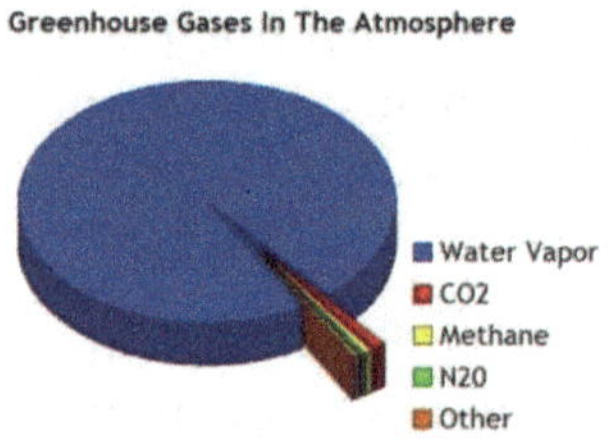

Water vapor is the main greenhouse gas.

The former vice president of the US, Kamala Harris, expressed her concern about how so many of our young people are struggling with "climate mental health" problems. "They worry about whether to start a family [in view of what's ahead]. They might have to get a job! The stress! How can they adapt the education they have now to their activism?"[18]This is an example of the effect such propaganda, amounting to terroristic threats, has had on our children, who are too young to be aware of all of the false predictions from decades ago that never came to be.

Obviously the goal of the climate alarmists is to raise a generation of young climate zealots who have been completely taken in by the invented climate scare, and who totally lack critical thinking skills and tolerance for alternative viewpoints. [19] And since these falsehoods have been delivered for nearly forty years, two generations have actually been infected. Those in power have done this quite easily by inserting this propaganda into all textbooks, even some that are not science textbooks.

Even college textbooks are contaminated with this disinformation. One physical science textbook contains the same incorrect pie graph of greenhouse gases in which the predominant one — water vapor — is missing. This book reads,

> "The large-scale disruptions of normal life that global warming seems to be on the way to bringing about are not likely to be met passively by those involved. It is entirely possible that within the lifetimes of many people alive today, tens of millions of Latin Amer-

> icans, hundreds of millions of Africans and a billion Asians will run short of freshwater and food. Rising seas made worse by storm surges will flood out hundreds of millions, eventually billions of people." [20]

Notice how cleverly this is written, saying, "global warming **seems to be on the way to bringing about**" and "It is entirely possible **within the lifetimes of many people alive today,**" etc., etc.

This same college textbook contains Michael Mann's "hockey stick" graph and another showing an exponential rise in carbon dioxide after the year 1955. The percentage of CO_2 has risen linearly, not exponentially. This book also neglected to mention that during the period of time in which carbon dioxide was rising, the earth's average temperature was actually **declining**.

Instilling fear is one of the most widely used techniques in climate change propaganda. Those perpetrating this disinformation seem to have no qualms about this fear campaign. Rupert Read, a co-founder of Extinction Rebellion, has been telling school children as young as nine years of age, "I don't want you to be calm. I want you to act like our house is on fire, because it is. People probably sometimes ask you, 'What are you going to be when you grow up?' But the question also has to be asked, 'What are you going to do **if** you grow up?'" Here he is implying that they may not live to adulthood because of climate change, and they've gotten the message. [21]

Children are even encouraged to turn on their parents if they express any skepticism about the official narrative.

"The National Center for Science Education (NCSE) recommends that parents and community members form alliances to ensure that science and only science is taught in our schools. The NCSE says it vigilantly monitors the accurate teaching of climate change and mobilizes local communities and educators to respond effectively when problems arise."[22]So any teacher who dares to question this dogma (as s/he should in any case) is in danger of being disciplined or even losing his/her job. Unfortunately, even teachers' unions have joined the ranks of climate alarmists.

Official UN UNICEF guidelines instruct teachers not to dismiss or try to minimize any worries their students have. Instead they affirm that children should be told that they have a right to be scared and that they should become radical climate change crusaders themselves.

Mass Formation Psychosis, a term introduced by Mattias Desmet to describe the reaction of large groups of people during the COVID pandemic to mass media influences such as TV and social media, in behaving like robots or zombies, adhering to all the rules on masks, vaccinations and mandates without thinking. As a result, the lockdowns and school closures were largely successful. These people believed they were acting responsibly, but they were either reacting out of fear or merely copying what others were doing. This is not a new phenomenon, a fact to which the witch hunts of the 16th century attest. There have been many other cases of Mass Formation Psychosis throughout history, as outlined in Charles Mackay's 1841 book, *Extraordinary Popular Delusions and the Madness of Crowds.*

The term "climate lockdown" is called a conspiracy theory. But it refers to a group of restrictions that have been threatened, such as fifteen-minute cities, travel restrictions, forced veganism and other drastic measures to be forced on the public in the name of reducing CO_2. This is openly discussed on the website climatecosmos.com, which points out:

> "The recent pandemic gave us a glimpse into how lockdowns could impact the environment. During COVID-19 lockdowns, many cities witnessed clearer skies, reduced air pollution and wildlife returning to urban areas. According to NASA nitrogen dioxide pollution in parts of China decreased by as much as 30% during the lockdowns. This unexpected environmental benefit has led some to consider whether similar measures could be used to tackle climate change more permanently." [23]

If ever implemented, such a policy would be a severe assault on personal freedoms and destroy economies.

Propaganda is also known by the gentler term, public relations. There are all kinds of courses and certifications in this field. A person can earn a degree in public relations, which is a subset of psychology. Some strategies used in this industry are: repetition of key messages, simplification of complex messages, and presentation of only one side of an argument to shape perceptions.

"Nudge theory" is a technique used to shape human behavior that is taught in these courses. It is a way to "influence the behavior and decision-making of groups or individuals. Nudging contrasts with other ways to achieve compliance, such as education, legislation or enforcement. . .It has influenced British and American politicians. Several [over 200] nudge units exist around the world at the national level as well as at the international level."[24]

The 2008 book, *Nudge: Improving Decisions About Health, Wealth and Happiness* by Cass Sunstein and Richard Thaler inspired powerful politicians such as Barack Obama and David Cameron to set up committees to incorporate these ideas into their policies. Some people are very concerned that such tactics could push the public into gradually accepting totalitarianism. [25]

Censorship and Other Measures

To prevent anything but their own testimony from getting any publicity, those in power who are perpetrating this giant scam have used several game plans. First, the Google search engine and others like it will serve up articles and comments friendly to the official climate narrative on the first page. They know that very few researchers bother to scroll past the first page.

Then we have the abuse of the peer review system. Peer review was originally set up to evaluate the quality, validity and originality of research before it is published to ensure high standards in scientific work. This is a good thing. However, this system has come to be corrupted by the "old boy/girl system". Scientists of Climategate fame were telling their buddies, "I'll peer review your paper if you peer review mine."

These Climategate scientists actively conspired to keep the work of anyone who disagreed with them out of scientific journals. There are several cases where journal editors were actually fired for publishing the work of climate realists. These actions served to intimidate other journal editors. It is no wonder that they were able to claim that many more published papers advanced the narrative of human caused climate change than those presenting alternative views.

On July 23, 2025, the Trump administration's Department of Energy produced a report, entitled "A Critical Review of Impacts of Greenhouse Gas Emissions on the U.S. Climate". This report was written by five well-known scientists known to be climate realists, the Climate Working Group CWG): Roy Spencer, John Christy, Judith Curry, Steven Koonin and Ross McKitrick.

It's interesting to see the headlines describing this report. A *New York Times* headline reads: "Energy Department Attacks Climate Science in Contentious Report". The AP report says, "Scientists give harsh grades to Trump administration work aimed at undoing a key climate finding." The headline on the Bulletin of Atomic Scientists website states, "Trump administration climate report a 'farce', scientists say," A rapidly composed "Factcheck" accused these scientists of over 100 errors in their 140-page report. This gives you an idea of how expertly the media can spin truth and smear climate realists.

Some Recent Fear-Inducing Climate Claims

"Climate change is officially the leading threat to imperiled species in the United States. For the first time scientists have identified climate change as the top driver of human-caused biodiversity loss among species protected under the Endangered Species Act." -Stephanie Waldek, April 24, 2025 NOTE: This false statement ignores the fact that offshore wind turbines have been proved 100% to be causing the deaths of hundreds of marine mammals — three species of whales plus dolphins, which were supposedly protected under this act. The companies responsible for these atrocities have been given official permission by NOAA to kill these animals.

"Wildfires are getting more intense around the world due to human-driven climate change. A new global wildfire report reveals how **human-caused warming** made fire weather dozens of times more likely." -Kenna-Hughes Castleberry, Nov. 6, 2025 NOTE: This statement runs counter to statistics, which anyone can look up. Further, it incorrectly assumes that the AGW hypothesis has been proved.

"Europe **could** get 42 more days of summer **by the year 2100** due to climate change." -Stephanie Waldek, Dec. 1, 2025 (emphasis added)

"How climate change could make Earth's **space junk** problem even worse – **Rising concentrations of greenhouse gases** decrease the atmosphere's ability to devour space junk, a new study finds." -Thereza Pultarova March 17, 2025

"How **unexploded bombs** cause environmental damage — and why **climate change exacerbates the problem.**" -Christina Greene and Sarah Njeri, Sept. 4, 2025

"Satellite data reveals 2023 was record breaking for marine heatwaves — are we at a climate 'tipping point'? The impacts ripple into human systems, reducing fishery yields, straining aquaculture and affecting industries that rely on stable ocean conditions." -Victoria Corless, July 24, 2025 NOTE: No data was presented to verify these claims.

"Is Earth 'on the brink'? 2024 was likely the **hottest year in 125,000 years.** Two new climate reports warn Earth's vital signs are flashing red after the hottest year in 125,000 years, but say recovery is still possible with swift action." -Sharmila Kuthunur, Oct. 29, 2025

"Climate models reveal how human activity **may be** locking the American Southwest into **permanent drought**: A new wave of climate research is sounding a stark warning. Human activity may be driving drought more intensely — and more directly — than previously understood." -Pedro DiNezio and Timothy Shanahan, 8/24/25

"Fall 2024 was Earth's **warmest autumn ever**. Fall 2024 makes history in the record books as the warmest on record in the past 130 years." -Meredith Garafalo Jan. 2, 2025

"October 2024 set multiple US records for the **driest month ever**. Drought conditions expanded to record-breaking levels across the US during the month of October as temperatures also climbed to record highs." -Meredith Garafalo, 11/21/24

"20 years of satellite data reveal **'staggering' levels of glacier melting, sea levels rising.** Every centimeter of sea level rise exposes another 2 million people to annual flooding somewhere on our planet." -Sharmila Kuthunur, 2/28/25 **NOTE: What is truly staggering is the level of dishonesty here.**

"Satellites find likely cause of mysterious African elephant deaths in 2020: **Climate extremes seem to have exacerbated toxins in the elephants' water supplies."** -Sharmila Kuthunur, 12/5/24

Such propaganda will likely continue on and on. Our only defense against it in our monumental David vs Goliath battle is

1. Checking claims to arrive at the truth,
2. Communicating the truth to everyone who will listen, and
3. Teaching our young people how to think critically and to be skeptical of everything they see on TV and social media.

One big advantage of reaching zero hour on the Climate Clock is that, if it is then too late to do anything about the situation, the propaganda barrage will hopefully end.

---

1. Fourth National Assessment,"Volume II: Impacts, Risks, and Adaptation in the United States," https://nca2018.globalchange.gov
2. "An Inconvenient Truth for Al Gore", Michael Dobbs, The Washington Post, 10/12/2007
3. Steven Mosher and Thomas Fuller, *The Crutape Letters,* 2010.
4. https:junkscience.com/2011/11/suzuki/where/will/santa/live
5. Dr. Patrick Moore, in a March 15, 2019, tweet.
6. Roasanna Xia, Sept. 11, 2024
7. Caroline Hickman, "Climate Anxiety in Children and Young People", *The Lancet Planetary Health,* December 2021.

8. Maggie Astor, "The 2017 Hurricane Season Really Is More Intense Than Normal." The New York Times, Sept. 19, 2017

9. Michael B. Sauter, "Cities Around the World that will Soon be Under Water," 247wallst.com, 4/7/22

10. Ibid.

11. Mark Hertsgaard, "Climate Denialism is Literally Killing Us," The Nation, Sept. 25-Oct. 2 issue

12. Philip Bump, "This Map of Sea Level Rise Is Probably Wrong. It's Too Optimistic," https://www.theatlantic.com/politics/archive/2013/11/map-sea-level-rise-probably-wrong-its-too-optimistic/354766/, Nov. 4, 2013

13. Robert Craychik, "Greenpeace Founder: Global Warming Hoax Pushed by Corrupt Scientists 'Hooked on Government Grants'", Breitbart News, March 7, 2019

14. Eric Worrall, "Damage Control: CNN Explains "Global Warming" Does Not Always Mean Warming,"https://www.sgtreport.com/2019/03/damage-control-cnn-explains-global-warming-does-not-always-mean-warming/, March 3, 2019

15. https://www.msn.com/en-us/weather/topstories/10-climate-tipping-points-the-world-needs-to-avoid-at-all-costs/ss-AAY5bWM?bk=1&ocid=msedgntp&cvid=14b514f58c8b40adb444b951031a1919

16. Paul Watkiss and Kit England, "Adaptation Finance and the Private Sector: opportunities and challenges for developing countries," *Climate Resilience Alliance*, Sept. 2025.

17. Greenhouse Gases, A Student's Guide to Global Climate Change, US EPA

18. Sean Hannity, video clip; Fox News, March 9, 2023

19. Tom Nelson, "Climate Brainwash", Gorilla Science YouTube Channel, November, 2025.

20. Konrad B. Krauskopf and Arthur Beiser, *The Physical Universe, 13th Edition*, McGraw Hill, 2010, p. 111

21. As articulated in his book, *Parents for a Future*

22. Ibid

23. The Climate Cosmos Team, "The Push for Climate Lockdowns: Is This the Next Step?", Dec. 3, 2025.

24. Wikipedia

25. "Nudge Theory: What Fifteen Years of Research Tells Us About Its Promises and Politics," *The Conversation*, Sept. 6, 2023

Chapter 9

Unsustainable "Solutions"

> "The right response to the nonproblem of global warming is to have the courage to do nothing." —Christopher Monckton

The quick-fix "solutions" to climate change offered by the elites running the world are wind power, solar power, and biofuels. These multibillion-dollar industries are now in full swing. They and their friends are cashing in on this scam to the tune of trillions of dollars. On the day he took office Joe Biden signed an executive order canceling the Keystone Pipeline and multiple oil drilling leases. This started a catastrophic rise in the price of gasoline, heating oil and diesel fuel in the US, causing significant inflation. Although the Trump administration has reversed many of these policies, reversing the resulting inflation will be more difficult.

Wind power is neither green nor sustainable.

You've been told repeatedly by the media — in magazines, TV, the Internet, etc.— that electricity produced by the wind industry is clean, essentially free energy, better for the environment than traditional sources such as coal and natural gas. None of these claims is true.

These wind installations injure, maim, and kill hundreds of thousands of birds and bats each year in clear violation of federal law. The Golden Gate Audubon Society in California reported that the wind farm at Altamont was killing 10,000 birds, including 1,123 birds of prey, each year. Due to a lawsuit, the Altamont wind turbines have had to cease operation for the four months annually that these birds are migrating.[1] Offshore wind turbines have similar impacts on marine birds, and, according to the Union of Concerned Scientists, wind farms located offshore also impact fish and other marine wildlife. [2] Strangely, the numbers and types of birds killed by wind turbines are ignored by wind farm proponents, even though they are substantial.[3] How can this be considered environmentalism?

Currently the construction of an offshore wind farm about 24 km (15 miles) off the coast of Massachusetts is underway. The foundation pieces for the huge wind turbines, called monopiles, are being driven into the sea floor by pile drivers. Pile-driving noise can deafen, injure or even kill marine mammals. At least fourteen dying humpback whales have recently been washed up on beaches in this area. The people building these projects are fully aware of the damage to marine life that they are causing and will cause. Although killing such creatures is illegal, these companies have been given special permission by the government to injure or kill up to hundreds of whales and thousands of other marine mammals. "It is hard to see how this hell can be allowed under present laws, such as the Marine Mammal Protection Act and the Endangered Species Act," noted David Wojick of the Heartland Institute.[4]

In May of 2022, Sean Hayes, a NOAA scientist, issued a warning that "the development of offshore wind poses risks to these species [whales]" and that "these risks occur at varying stages, including construction and development, and include increased noise, vessel traffic, habitat modifications, water withdrawals associated with certain substations."[5]

Steve Gray, a former congressional candidate, wondered, " Where are the environmentalists? No outrage. No protests against a Democrat led Government backing offshore windmills. The hypocrisy is deafening. 14 dead whales and counting."[6]

Amazingly, Greenpeace, an "environmentalist" organization that formerly fought for whales and other creatures, now asserts that there is no scientific evidence linking the construction of offshore wind turbines to the deaths of right whales. They argue that the primary threat to whales is climate change (believe it or not!), collisions with boats and entanglement with fishing gear. They maintain that transitioning to renewable energy like offshore wind installations is essential to prevent climate change, which they call the greatest threat to marine mammals.

Investigating this irony more closely, Lisa Linowes of the Save Right Whales Coalition discovered that Greenpeace and other organizations purporting to be protecting wildlife have taken almost five million dollars from the multibillion-dollar wind industry. This is a betrayal of the very principles on which these environmental groups were founded and demonstrates that they can all be bought for a price.

The Endangered Species Act, passed in 1973, used the Precautionary Principle that any activity that risks the extinction of a species should be illegal. But now NOAA has given the wind industry at least eleven permits to "harass" marine animals, including the North Atlantic right whale. It is believed that fewer than 340 of these creatures remain. NOAA continues to insist there is no connection between wind installations and the deaths of whales. In 2023 Rutgers Professor Gerasoulis, a computer science expert, built a software program that investigates these events and has concluded with 100% certainty that these offshore wind operations are definitely killing the whales. [7]

Recently the Trump Administration paused all large wind power projects off the east coast of the United States because of national security concerns. This decision affects five projects under construction in that area. The Department of the Interior discovered that these wind turbines were interfering with radar and were thus threatening the nation's defense. Radar "clutter" can obscure all moving targets, creating false signals. The administrators of these projects have already filed lawsuits against the US government.

Hundreds of square miles of forests have been clear cut to provide space for wind farms, causing extensive environmental damage. Fifteen million trees were cut down in Scotland to make way for wind turbines. It's been estimated that wind and solar installations require 400 – 1000 times the amount of land to produce the same amount of power as conventional power plants. There are land use disputes going on all over the world.

Modern wind turbines depend on rare earth minerals mined primarily in China.[8] China has a poor record of environmental stewardship. So the process of extracting these minerals imposes wretched environmental and public health conditions on China's people.

A typical modern wind turbine, over 152 meters (50 stories) tall, the height of the Washington Monument, contains more than 8,000 different parts, made from steel, fiberglass, cast iron, and concrete. It takes thousands of tons of coal to make just one wind turbine. There is no other way to make the steel and concrete. Each blade, made from fiberglass and plastic, is the length of a football field. Because these blades can't be recycled, they are cut up and buried in large fields. A small wind farm requires the same amount of nonrecyclable plastic as all of the plastic straws on Earth. Plastic straws are recyclable.

One of these turbines has a power rating of about 2 megawatts (million watts). Because the wind isn't always blowing at the right speed, these are only about 20–30 percent efficient. On good days this provides power for 20,000 homes. In contrast, a gas turbine generator with the same power rating takes up the space of one tractor trailer container.

One important component of a wind turbine is a magnetic material made from neodymium and dysprosium, rare earth minerals mined almost exclusively in China. Since rare earth metals are quite toxic, they are no longer mined in the United States.

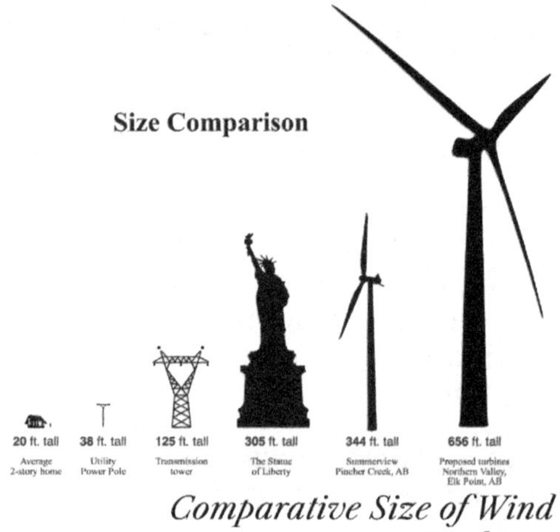

Comparative Size of Wind Turbines

In the Chinese village of Dalahai, where the dumping of waste materials from these occurred, villagers' teeth had begun falling out, their hair turned white at unusually young ages and they suffered from severe skin and respiratory diseases. Children were born with soft bones. Abnormally high rates of cancer, osteoporosis and skin and respiratory diseases were reported. The nearby lake's radiation levels are ten times higher than in the surrounding countryside.[9]

Considering the number of animals these monstrosities kill and the amount of fossil-fuel-generated energy required to mine the materials needed to build them, how could anyone who truly cares about the environment approve of their use?

To be competitive with other sources of power, wind farms require huge government subsidies paid for by taxpayers. They also need vast stretches of land. Rockefeller University professor Jesse Ausubel has calculated that meeting New York City's electricity needs alone would require blanketing the entire state of Connecticut with these towering turbines.[10]

Because the wind doesn't blow all the time, every wind turbine must be backed up by expensive coal or gas power plants that sit idle when the wind is blowing at just the right speeds. These conventional sources must then kick in whenever the wind dies down in order to prevent factories, schools, offices and homes from experiencing blackouts or brownouts.

Richard S. Courtney explained,

> "Windfarms for power generation provide intermittent power, so they merely displace thermal ["fossil fuel"] power stations onto standby mode or to operate at reduced efficiency while the thermal power stations wait for the wind to change. They make no significant reduction to pollution, because thermal power stations must continue to use their fuel and produce their emissions while operating in standby mode with reduced efficiency. Their emissions from operating at a low output are actually increased."[11]

Germany recently destroyed its last nuclear plants, forcing utilities to buy "renewable" energy at several times the cost of conventional power. In the United Kingdom, *The Telegraph* reported that rather than providing cheaper energy, wind power costs the electric companies £50 per megawatt-hour, compared to £15 for conventional power - over three times the cost.

The same is true of any electrical grid. Wind turbines respond only to the wind, so their contribution to the supply of electricity on the grid is essentially inconsistent and unreliable. The wind may not be blowing when demand is high. When the demand is low, it could be blowing just right. Wind turbines don't operate when the wind speed is below six meters per second and must be shut down if it is greater than 25 meters per second (to protect blades and bearings).

This is the dirty little secret behind why Big Oil loves wind power. Coal or natural gas generating plants will always be needed to back them up. People who operate power plants tell us that continually powering up and down is much more wasteful of energy than constant operation.

Broken wind turbine

In February, 2019, John Hinderaker warned, "Here in Minnesota, we are enduring a brutal stretch of weather. The temperature hasn't gotten above zero in the last three days, with lows approaching -30°. And that is in the Twin Cities, in the southern part of the state. Yesterday central Minnesota experienced a natural gas "brownout," as Xcel Energy advised customers to turn thermostats down to 60 degrees and avoid using hot water. Xcel put up some customers in hotels. Why? Because the wind wasn't blowing."[12]

Professor Randy Simmons added,

> "Over the past 35 years, wind energy – which supplied just 4.4% of U.S. electricity in 2014 – has received $30 billion in federal subsidies and grants. These subsidies prevent people from learning the uncomfortable truth of just how much wind power actually costs. They transfer money paid by taxpayers [without their knowledge] to wealthy wind farm corporations, many of which are units of foreign companies."[13]

Several studies have placed doubt on the value of wind turbines. A 2011 study at Purdue University and a 2016 study in Scotland found that wind farms were actually increasing surface temperatures. A 2018 study concluded that wind power in the US could warm surface temperatures by 0.24° C., nearly one-fourth of the global temperature increase experienced since 1800.[14]

The Institute for Energy Research reported that the actual cost of wind energy is about twice that estimated by the Department of Energy. Because the useful life of a wind turbine is only about 10-15 years, if they are not properly maintained, rather than the optimistic 30-year figure provided by the government, claims by the wind industry that wind power is becoming cost competitive with fossil fuel-generated power are absolutely false.[15] The maintenance cost for a 2-MW wind turbine costs anywhere from $40,000–$200,000 per year.

If wind power was less expensive than — or even merely cost-competitive with — coal and natural gas energy, these subsidies and mandates would be unneces sary.[16]

Added to these problems are the adverse health effects of living anywhere near these wind turbines. The incessant grinding noises they make have been known to cause tinnitus [ringing in the ears], vertigo, panic attacks, migraine headaches, sleep deprivation, and even heart disease, according to groundbreaking research published by Dr. Nina Pierpont, a leading New York pediatrician.

Dr. Pierpont studied the symptoms of people living near wind turbines in the United States, the United Kingdom, Italy, Ireland and Canada. Her findings have led her to confirm what she has identified as a new health risk: wind turbine syndrome (WTS).[17] This is the disruption or abnormal stimulation of the inner ear's vestibular system caused by turbine infrasound (low-frequency noise), the most distinctive feature of which is a group of symptoms which she calls visceral vibratory vestibular disturbance, or VVVD.

Wind turbines kill millions of birds and bats.

They cause problems ranging from internal pulsation, quivering, nervousness, fear, a compulsion to flee, chest tightness and tachycardia (increased heart rate). Turbine noise can also trigger nightmares and other disorders in children and harm cognitive development in the young.[18] If wind turbines are affecting people this way, what are they doing to farm animals and wildlife?

Locating wind farms offshore or in unpopulated areas is not a practical solution. This is because electricity must then be transported long distances to reach the cities, where the demand is greatest. The farther electricity has to travel from the producer to the user the more energy is dissipated en route, wasting power.

Offshore Wind Turbines

Winds are stronger and more constant in offshore sites than on land installations. However, expensive mountings must then be built to prevent large ocean waves from dislodging them from their moorings.

In February of 2021 Texas and nearby areas were hit with the worst winter storm in over a century, with several consecutive days of freezing temperatures. Hundreds of wind turbines iced up and failed to operate, causing a nightmare for ERCOT, the electricity provider. Millions of people were without electricity for several days. An estimated 200 people died. The cost of these unreliable wind turbines was over thirty billion dollars: over $1,000 for every man, woman and child in Texas.

In view of the above, the massive subsidies of around 40% of the cost of each wind turbine provided to wind power companies in the US have proved to be a tremendous waste of tax dollars. Industrial wind farms are ridiculously expensive and destructive to communities and local environments. Fragile ecosystems are destroyed. Birds and bats are killed. Animal habitats are disrupted, and wildlife is driven from the area. Noise and vibrations cause human health problems, including sleep problems, headaches and nervous disorders. Home values and real estate prices are reduced, with some properties becoming impossible to sell. Energy from wind power is not a viable alternative to carbon-based energy production.

The Truth About Solar Energy

Photovoltaic cells, made of semiconducting materials such as silicon, convert sunlight directly into electricity. They are useful for all sorts of jobs such as powering watches and calculators and road signs that light up. When photons of sunlight strike these semiconducting materials, they knock electrons loose from their atoms. These electrons are then directed through the semiconducting material to a light or motor.

Under ideal conditions, when there is a cloudless sky at noon, at an angle of 90 degrees to the direction of sunlight streaming in from the sun, one square meter of the panel will produce about 1000 watts, or one kilowatt, of power. This is known as Watt Peak.

But often it is rainy or cloudy. Only when the sun is highest in the sky at noon on a cloudless day will you have ideal conditions. The solar panel must face south in the northern hemisphere and north in the southern hemisphere. It must be at the same angle as the latitude of the installation. During winter months, when the sun doesn't get as high in the sky, and during morning and evening hours, the wattage — and the efficiency — of the photovoltaic panel drops off. And then, of course, it is nonexistent at night when the sun isn't shining. This is why a typical solar panel has an average efficiency of only about one-sixth (17%) of its peak rated power. And the efficiency becomes even lower over time as the solar panel starts to wear out.

The average cost of a 6.7 kW residential solar panel installation in 2021 was $17,800. Applying the tax credit paid for by taxpayers, you'd save about 26% of this cost. So you'd actually pay about $13,170. This works out to between $2.75 and $3.35 per watt, plus electrical and roof work, trenching, batteries, etc. Through our taxes, we are helping others finance the construction of their solar projects. Unfortunately people who are too poor to afford their own home are helping richer people to purchase these systems.

A YouTube video presenter claimed that his system, rated at 4800 watts, was providing 4800 watts of electrical power. He used this figure to compute the time needed to recover his original expenditure as only 7.1 years. This is a mistake many people make, because they don't understand that the efficiency of such a system will never exceed about 17% of the rated power (peak wattage). He will be lucky to get 816 watts (17% of 4800 watts). The time needed to recover the cost of this system will thus be about 42 years, not 7.1 years. The average life span of such a system before it has to be scrapped is about 21 years. So even with the 30% tax credit a person will never recover the investment s/he made in it. And this doesn't include the substantial cost of having it scrapped after it wears out. So solar power is not "free", as advertised. In fact, consumers investing in this technology gets back a return of about 10 cents for every dollar spent. [19]

Commercial solar projects consume a tremendous amount of land area. Solar plants need about 400 - 1000 times as much land as natural gas or nuclear electrical generators for the same amount of power. These huge solar farms impinge on and threaten natural ecosystems. How could any real environmentalist approve of such systems?

One type of commercial solar electricity generating plant is a concentrated solar power plant, also known as solar thermal. In this type of arrangement each of an array of thousands of mirrors surrounding a receiving tower is positioned at the precise angles needed to reflect sunlight onto the tower. The Gemasolar power plant in Spain consists of 2,650 such mirrors and occupies 185 hectares (457 acres) of land. This produces enough power to supply 25,000 homes. Sierra Suntower, a plant in the Mojave Desert in California, has two towers and 24,000 mirrors, is rated for 5 MW (million watts) providing energy for 4,000 homes.

You can learn more about these solar projects by accessing solarpaces.nrel.gov, which provides a list of projects, their cost to construct, capacity, location and current operational status. Because the sun's position in the sky is constantly changing, due to Earth's rotation and changing seasons, each of the mirrors depicted here must have its own motor to change its position in order to keep the reflected sunlight focused on the tower. The central tower contains water or molten salt at a temperature of about 500°C. circulating through pipes. This molten salt heats water to steam, powering an electrical generator.

A thermal solar farm.

The advantage of this type of arrangement is that some of the hot molten salt can be kept in an insulated tank to be used for heating water at night or on cloudy days. This type of operation, called a thermal battery ["thermal" meaning "heat"], doesn't need a battery backup for times when the sun isn't shining as do traditional photovoltaic plants.

These plants are extremely expensive. They take up a tremendous amount of acreage and destroy wildlife habitats. Maintenance costs to keep such a plant running are high, because each mirror must be at peak reflectivity and exactly the right angle. Solar plants are often located in deserts where cloudy days are rare. But two problems arise from this. First, water, scarce in a desert, is required to operate such a system. Deserts are far from many population centers. A great deal of electric power is lost in transporting it over long distances.

Feeding solar energy into the electrical grid is tricky. Peak solar generation occurs in the early afternoon when the sun is highest in the sky. Peak demand for electricity doesn't occur until the evening hours. This mismatch between peak output and peak demand has been dubbed the "duck curve". In the evening, as solar generation decreases due to the sun's position in the sky, system operators must use coal, natural gas or nuclear backup generation systems to prevent power outages and brownouts. Experience has shown that if the input to the grid from a solar system exceeds about 6%, there is actually a negative effect on the power grid that overloads it.[20]In this case production must be curtailed. This electrical energy cannot be stored in batteries, because battery technology hasn't developed to the point where batteries can hold more than a few hours of electrical power.

In 2016 California ISO (Independent System Operator) was forced to curtail over 3.08 GWh (billion watt hours) of electricity, which was wasted because supply exceeded demand. As ISO put it,

> "While curtailments are currently representing only a fraction of overall load, the issue is expected to intensify in the coming years as the state strives to achieve its 50-percent renewable target. . .Curtailing renewables is counterintuitive to California's environmental & economic goals. It reduces the output from the renewable plants

> in which the state has invested, and could result in **overbuilding** renewable plants to ensure that the state meets its 50% renewable mandate. **Overbuilding** the state's electrical system isn't financially sound." [21](Emphasis added)

Large solar projects receive government subsidies of at least 30% of their cost, paid for by the taxpayer. Many taxpayers have no idea that they are actually footing a substantial part of the bill for these projects.

Many states are now imposing penalties on power companies that fail to include a specified percentage of the power they produce from these expensive wind and solar sources.[22] There is a double whammy here. First an additional burden is put on taxpayers to pay the subsidies. Second, because wind and solar power systems are so expensive, the power companies have no other option but to pass this increased cost of electricity onto their customers. The cost of electricity has more than doubled in Great Britain, Germany, California and many other places.

The construction of photovoltaic panels also causes great harm to the environment and human rights. An estimated 90% of the World's photovoltaics are produced in China, much of it with slave labor in Uighur Muslim concentration camps. The New York Times reported in 2014, "Although China may be a cheaper place than Europe for producing solar panels, the savings come at a higher cost to the environment." In their analysis scientists added together all the costs involved in production: the mining of raw materials, transportation, and the power requirements for production of these panels.

"Sometimes the environmental costs of solar panel production can be lost among the drive to encourage the development of clean energy," declared Huang Xianjin, a Nanjing University professor who studies land use. Huang and two other professors submitted a letter to the journal, *Nature*, arguing that China needs to take significant action to offset the environmental damage the solar industry is causing there.[23]

A solar panel manufacturer in Zhejiang Province halted its operation after local residents complained about serious air and water pollution caused by their production. The Luoyang Zhonggui High-Technology Company produces a highly toxic byproduct of polysilicon used in solar panels called silicon tetrachloride. "The land where you dump or bury it will be infertile. No grass or trees will grow in the place...It is like dynamite - it is poisonous, it is polluting. Human beings can never touch it," declared Ren Bingyan, professor at Hebei Industrial University in China. The irony is that solar is touted as green energy.[24] Because of the increasing demand for solar energy, the price of polysilicon soared from $20 to $300 per kilogram in five years. Dozens of new polysilicon plants are riding the crest of China's prosperity. The combined capacity of these new factories is estimated at 72 million kg (158 million lbs), over twice the amount of polysilicon produced in the rest of the world.

Then there is the problem of the disposal of worn out solar panels. Michael Shellenberger expressed his concern about environmental hazards resulting from this practice. Typically they contain cadmium, lead and other toxic chemicals which can't be extracted without taking the whole panel apart.[25] Such panels left in landfills break apart and release toxic waste into the ground or bodies of water.

In a 2016 study Electric Power Research Institute concluded, "Solar panel disposal in "regular landfills [is] not recommended in case modules break and toxic materials leach into the soil." Over a period of time rainwater will wash cadmium out of solar panels and into the environment. This fact was brought up when local people In Virginia rejected a proposal to construct a 6,350 acre (2,570 hectare) solar farm in Spotsylvania County. Sean Fogarty of Concerned Citizens of Fawn Lake stated, "We estimate there are 100,000 pounds [45,000 kg] of cadmium contained in the 1.8 million panels."[26]

There have been some promising advances in solar electricity generation. But many more years of research and development are needed to make solar energy cost competitive and environmentally safe. In order to eliminate the need for backup by coal and natural gas, batteries, which are now totally inadequate, must be improved.

At this time the battery technology needed to store this energy is woefully inadequate. "The reason winter is cold is because of a lack of solar energy. The sun is low in the sky, days are short, and it is cloudy much of the time. Yet climate alarmists want people to be dependent on solar energy being generated under those conditions for their survival. They imagine that there is some storage technology which can store huge amounts of energy for long periods of time when the sun isn't shining or the wind isn't blowing."[27] The problem is that these haven't been found yet.

The Green New Deal and other carbon tax schemes anticipate having all of these problems solved within a decade or so. This is an impossible pipe dream entertained only by uninformed politicians and others who don't have a clue about the science or the technology.

Seven American refineries, which formerly processed 806,000 barrels of petroleum per day, have gone offline. Some simply shut down while others are being converted to biofuels like ethanol to meet the demands of the environmental lobby. "These losses have left the US with 124 operating refineries. That's 10% fewer refineries for gasoline, diesel and jet fuel than in 2016, even as demand has continued to increase."[28]

Energy blackouts are predicted for the near future because of the installation of "renewables" and the cancellation of dependable power sources. Wind and solar energy farms require either massive, utility scale battery farms or backup from reliable hydrocarbon-powered generators, which costs a fortune. This has been compared to a person jumping out of a plane and then wondering if he is wearing his parachute.

Environmental Destruction

It is difficult to fathom the extent of the environmental destruction caused by climate change mitigation. The full extent will probably not be known for several decades when people have come to their senses.

First of all is the sheer amount of land required for the placement of photovoltaic panels and wind turbines. A solar farm rated for one gigawatt of energy typically requires up to 12,150 hectares (30,000 acres) of land. In contrast, a natural gas plant producing the same amount of power can be built on less than 30 hectares (75 acres) of land, using 400 times less land. Farmland has been converted and forests cut down to make way for them. Such projects strip wild animals' homes from them, putting many species in danger of extinction.

These disrupt ecosystems by forcing wildlife to relocate and reducing the availability of food and shelter. They destroy nesting sites for many species of birds and bats. The reflective surfaces of solar panels can confuse birds, causing them to collide with the panels or abandon their nests. Solar farms disrupt food chains by altering the availability of prey or changing the behavior of predators. Solar panels attract pollinators such as bees and butterflies, which are essential for plant reproduction. They block access to flowers and sources of nectar. The heat solar panels generate can even alter microclimates, affecting the timing of plant flowering and pollinator activity. [29]

Their installation can alter drainage patterns, leading to increased runoff of water and soil erosion. When solar panels are damaged by hail, lightning or tornadoes or when they simply wear out after ten or fifteen years, the toxic materials — such as lead and cadmium — that they contain will leach out into soil and water tables unless they are properly recycled. This further amplifies their environmental footprint.

The construction of solar panels, almost 90% of which are manufactured in China, also imposes many environmental hazards. Several minerals, such as silicon, silver and rare earth metals like neodymium and dysprosium, are needed. The mining of these metals results in habitat destruction and loss of diversity.

There are several lawsuits pending to prevent further destruction of marine wildlife by offshore wind installations. It is estimated that only about 340 right whales are still alive. This species is in serious danger of extinction. Tragically, the companies installing offshore wind turbines have been given permission by the US government to violate the Endangered Species Act.

In July of 2024 a wind turbine blade — nearly the length of a football field — malfunctioned and splintered, dropping thousands of fiberglass shards into the ocean. These shards washed up onto several beaches on the Rhode Island coast. [30] Although people are diligently picking them up, it may be decades before these beaches are again safe for swimmers and people walking in their bare feet.

Another shattered wind turbine blade in the Midwest deposited numerous shards of fiberglass in a farmer's field, ruining his crop. This is just the beginning. We can expect much more of this to happen in the future, because wind turbine blades have a limited lifespan. [31]

Offshore Wind Turbine Blade Debris

Global warming started as an environmental issue, and all of the major environmental activist groups quickly got on board with it. Yet, ironically these "environmentalists" quickly turned a blind eye to the destruction caused by the climate mitigation activities.

Grossly Over-Rated Biofuels

There's nothing new about biofuels, or fuels derived from plants such as trees and corn. They've been used since the dawn of history, when people started burning them for cooking and heating their homes. Later, with the mining of coal and discovery of petroleum in the late 1800s, biofuels lost their popularity, especially in developed nations.

In 2020 global biofuel production reached 147 billion liters (39 billion gallons) with the United States and Brazil being the largest producers.[32]According tho the International Energy Agency (IEA), biofuel consumption amounted to about 3% of renewable energy, most (ethanol and biodiesel) being used for transportation. The production of biofuels derived from food crops competes with land used for food production. This drives up the cost of food and contributes to global food insecurity.[33]

Ethanol Storage Tanks

One type of biofuel, ethanol (ethyl alcohol), can be made from either corn or cane sugar. This is the same kind of alcohol as is found in alcoholic drinks. During the two world wars in the first half of the 20th century, biofuels were made from several different substances, including potatoes. But when huge supplies of crude oil became available in Texas and Oklahoma, gasoline became a less expensive alternative to biofuels.

In the 1970s the Club of Rome told us we'd soon run out of oil. But Michael J. Economides, one of the world's foremost experts on the geopolitics of energy, reported, "Approximately 6-8 trillion barrels of oil from conventional resources, with the same or more from unconventional [tar sands, shale oil etc.] resources are believed to be 'in-place' and as yet untapped," with a total of 12-16 trillion barrels in place.[34] But since the world started using oil, only about one trillion barrels of petroleum have been used. So the big question is: with all these available petroleum resources, why is there a need to use valuable agricultural land and rob the earth of virgin prairie and tropical rain forests to produce ethanol?

A group of Democrats led by Congressman Henry Waxman made a deal with the corn lobby. In exchange for its financial support of their political campaigns, they engineered a mandate to blend corn-derived ethanol with gasoline as an amendment to the 1990 Clean Air Act.[35] Advertised as another great source of alternative, clean energy to solve the world's energy problems, ethanol started being added to gasoline in amounts up to 15% without the consent of the consumer. It is now difficult in many parts of the United States to find gasoline stations that offer the option of pure gasoline without ethanol.

But gasoline containing ethanol quickly destroys boat and lawn mower engines. Many consumers have had the unfortunate experience of their lawn mowers and rototillers being destroyed because they were never warned about the damage the gasoline/ ethanol mixture causes in these small engines. It damages rubber and plastic components of fuel systems, especially in older models built of materials that aren't resistant to this type of damage. This includes gaskets, which become brittle, two-cycle carburetors and primer and return lines, which start leaking after a while.

If ethanol isn't burned within sixty days, it will literally decay.[36] This is because ethanol fuels are hygroscopic. They attract water. Because of this, fuel separation can occur and damage the engine's carburetor. Fuel which has decayed in this way contains varnish and sludge deposits. Fuel that is 15% ethanol, called E85, cannot be used in a vehicle unless its engine has been specially modified.

Ethanol is now a worldwide phenomenon. With biofuel mandates now in over 60 nations, the competition between ethanol and food for agricultural land has become a moral issue. Groups like Oxfam and the Environmental Working Group oppose biofuels, because they create food shortages, raising the price of food.

Ethanol production is a huge business. Millions of hectares (acres) of virgin prairie and wetlands have been plowed up for the first time ever. The increased use of nitrate fertilizers in growing these crops has caused massive runoffs of nitrates into creeks and rivers, contaminating water supplies. Cleaning up this water has become very costly. Over a third of all the corn produced in the United States goes into the production of ethanol.[37] 1.7 million hectares (4.2 million acres) of additional land, an area double the size of Yellowstone Park, is now being used for biofuels. Native prairie never before used for agriculture has been tilled up. In 2021 57 billion liters (15 billion gallons) of ethanol were produced in the US.[38] Yet at the same time almost eight hundred million people in the world are malnourished. And with the war in Ukraine, food shortages are expected.

"The UN's Food and Agriculture Organization (FAO) reported a 45% increase in the world food price index during the 2008 food crisis. Wheat prices increased by 130% over 2007 levels. Similarly, soy prices went up by 87%, rice prices by 74%, and maize prices by 31%." Causes of this crisis included the growing demand for biofuels.[39] Now lobbyists for this industry are pushing for legislation that would raise the maximum percentage of ethanol allowed in gasoline from 15% to 30%.[40]

The University of Minnesota and the University of Wisconsin jointly studied changes in land use across the US resulting from the ethanol industry. The leader of this study, Chris Wright, expressed his concern that native grassland is being converted to agricultural land, which has a negative impact on the environment. The Environmental Protection Agency (EPA) noted in 2011 that this practice causes increased soil erosion and loss of wildlife habitat. These are unfortunate and unintended consequences of ethanol production.

3.8 liters (1 gallon) of gasoline delivers 123 million J of energy, whereas 3.8 liters (1 gallon) of ethanol delivers only 80.2 million J of energy, about 2/3 the energy of gasoline. So a vehicle gets fewer miles per gallon of gasoline/ethanol blend than it does on pure gasoline. It costs $3.95 US to produce 3.8 liters (one gallon) of ethanol from corn.

About 138 million joules of energy are needed to make one gallon (3.8 L) of ethanol. "Put another way," Professor David Pimentel of Cornell University stated, "about 70 percent **more** energy is required to produce ethanol than the energy that ethanol contains. Every time you make 1 gallon of ethanol, there is a net energy loss of 57 million joules."[41] He also reported, "The use of corn for ethanol has led to major increases in the price of U.S. beef, chicken, pork, eggs, bread, cereals and milk – a boon to agribusiness and a bane to consumers." This industry would not be able to compete with hydrocarbons were it not for the subsidies.

The Director-General of the UN Food and Agriculture Organization, Jacques Diouf, reports that using 10 kg (22 lbs) of corn to produce 3.8 L (1 gallon) of ethanol is already causing food shortages.[42]Government subsidies for the corn industry include tax breaks, grants, loans, and loan guarantees. This subsidy averages $1.65 US per gallon, coming right out of your taxes.

> "It is now clear that the federal corn ethanol mandate has driven up food prices, strained agricultural markets, increased competition for arable land and promoted conversion of uncultivated land to grow crops. In addition, previous estimates have dramatically underestimated corn ethanol's greenhouse gas emissions by failing to account for changes in land use."[43]

It is immoral to divert agricultural land that could be growing food for fuel. The revolution in Egypt called "Arab Spring" started with riots caused by food shortages that were directly caused by this newly instituted practice.[44] But then, believe it or not, "human-caused climate change" was blamed for these rising food prices. Jean Ziegler commented, "It is a crime against humanity to convert agricultural productive soil into soil which produces foodstuff that will be burned as biofuel."[45]

"We thought and hoped it would be a climate solution and reduce and replace our reliance on gasoline," commented Tyler Lark, a researcher with the Nelson Institute for Environmental Studies at the University of Wisconsin-Madison, and lead author of the study. "It turns out to be no better for the climate than the gasoline it aims to replace and comes with all kinds of other impacts." Dr. Lark stated that ethanol puts 24% more carbon dioxide into the environment for the same amount of power than gasoline does. [46]

Biodiesel fuel is a blend of about 80% petroleum diesel and 20% biodiesel, which is produced by treating soybean or another oil with methanol and a catalyst. This industry receives substantial government subsidies through the Biomass Crop Assistance program (BCAP). Producers are reimbursed up to 50% of their cost with payments extending up to five years for herbaceous feedstock and 15 years for woody plants.

Section 9003 of the Biorefinery Assistance Program guarantees loans to producers for the construction, development, and retrofitting of commercial-scale biorefineries.[47] Biodiesel is competitive only through these subsidies, which bilk the taxpayer. Since there is a surplus of petroleum and natural gas, ethanol and other biofuels are unnecessary.

Climate alarmists have been concerned about a drought in the western United States, for which they blame climate change. Yet 20%-30% of the water in the Colorado River, west of the Continental Divide, is extracted for growing corn for ethanol production. This is how the climate change scam contributes to the loss of water in this region.[48]In California fresh water is being dumped into the Pacific ocean. "Since the beginning of January [2023] a series of atmospheric rivers has disgorged trillions of liters (gallons) of much-needed moisture across drought-stricken California, but only a small fraction of that water has so far made it into storage. In the delta — the heart of the state's vast water system — nearly 95% of incoming water - trillions of liters – has been allowed to flow into the Pacific Ocean, according to data from the U.S. Bureau of Reclamation."[49]

Using Wood to Produce Electricity

In some states forests are now being clear cut, and the trees ground up and burned to produce energy. They are replanted with just one or two species of trees, reducing biodiversity from 10-20 species to just one or two species of trees. This practice disrupts nature's balance. Forests owned by the state of Massachusetts "are being logged at drastic, nearly clear-cut levels and at unsustainable rates."[50]

A clearcut forest waiting to be shipped

> "Biomass is an appealing 'easy out' for wasteful energy policies and land barons hungry for quick cash. This sweetheart deal for big landholders is being sold as 'carbon neutral', even though burned

> wood releases 1.5 times as much CO_2 as coal per unit of mass. Taxpayers are lured in by promises of cheap energy and reduced taxes on the myth that this practice is sustainable or healthy for forests, people or the climate."[51]

To comply with new regulations calling for the use of "renewables" in the United Kingdom, Drax, its largest electrical generation plant, is now burning wood shipped 4800 km (3000 miles) from forests in North Carolina. Under European Union rules, wood imported from outside the EU is considered to be a renewable, zero-carbon fuel source. Going along with that, the United States Congress declared nearly all forest biomass in the United States to be carbon free. But declaring a substance "carbon free" doesn't make it carbon free.

Burning wood releases several toxic organic compounds, including some that produce acid rain. Harmful emissions include carbon monoxide (a deadly gas not to be confused with life-giving carbon dioxide) formaldehyde, benzene and dioxins, all of which are carcinogens. Some particulates released have been known to cause asthma and other respiratory diseases as well as cancer.

When these biofuels are burned, just as with gasoline and coal, carbon dioxide is emitted.[52] Biofuel advocates claim that the combustion of these trees is "carbon neutral" since they absorbed carbon dioxide while they were growing. But it takes 40 to 70 years to replace a forest. If, as they say, there are fewer than ten years left to save the planet, the newly planted trees wouldn't even have time to reach maturity.

Even worse, when these trees are cut down, the soil microbes located underneath the cut trees release just as much carbon dioxide into the atmosphere as does the combustion of the trees themselves. So when a tree is cut down and burned, nearly **three times** as much carbon dioxide is released into the atmosphere as an equivalent amount of coal.[53]

In the "green" bureaucrats' flawed energy accounting system, the practice of burning wood on a large scale is considered superior to the combustion of coal that could have been mined right in the UK. The energy required to cut down this forest, convert the wood into pellets and then transport the huge quantity needed to run this immense facility has obviously never been considered. Yet this practice has been hailed as a great achievement by people who think they are green.[54]

In setting the goal of doubling Europe's 2015 renewable energy levels by 2030, the unelected bosses of the European Union decreed that cutting down forests to be burned in electric power plants instead of using coal or natural gas qualifies as zero-carbon, renewable energy. They did this after nearly 800 scientists warned them that this policy would actually increase CO_2 emissions.[55] The Renewable Energy Directive (RED) is now finalized. It is obscene that a huge quantity of wood – as much as all the other uses of wood worldwide — will now, due to such brainless policy decisions, be required to meet just a tiny fraction of Europe's energy requirements. How could this practice ever be construed as helping the environment?

Wood burned in an electric power plant emits up to three times as much CO_2 as natural gas. This is because of wood's carbon bonds, water content, and the lower temperatures at which wood burns. It's been estimated that replacing coal and natural gas with wood could result in CO_2 emissions 2-3 times greater than carbon-based fuels by the year 2050.[56] After the forest is cut down, the trees will no longer be there to absorb the CO_2 they took in from the atmosphere when they were alive. Whether or not the extra carbon dioxide will affect the climate, this regressive action demonstrates a significant flaw in the knowledge and judgment of bureaucrats making such foolish decisions.

This harebrained decision by the European Union sent a disastrous signal to the rest of the world. At one of the most recent COP conferences countries with tropical rain forests - including Indonesia and Brazil - jointly stated their goals “to increase the use of wood...to generate energy as part of efforts to limit climate change.”

So just what do they mean by "sustainable energy"? It certainly couldn't be photovoltaic panels. It couldn't be wind turbines or biofuels. The only alternative energy source that is dependable, cost competitive and — due to its small land requirements — environmentally friendly is nuclear energy. Nuclear energy **is** sustainable energy. But the climate activists want no part of it.

---

1. Jonathan Benson, “‘Green’ Debacle, NaturalNews.com, Nov 24, 2011
2. /www.ucsusa.org/clean-energy/renewable-energy/environmental-impacts-wind-power#.WqAYMM9x0xA
3. "Wind Farms Raise Temperatures at the Surface Level", Institute of Energy Research, March 20, 2023
4. David Wojick, "Whale Hell Looms in Massachusetts", Feb. 18, 2023, The Heartland Institute
5. Peter Partoll, Spate of Whale Strandings Near Offshore Wind Projects Becoming Impossible to Ignore – But Feds Still Deny Link, The Western Journal, March 4, 2022
6. May 3, 2023
7. Donna Anderson, "Professor Makes Stunning Discovery 'Absolutely 100% Offshore Wind Kills Whales'" *Climate Change Dispatch*, July 15, 2024.
8. http://instituteforenergyresearch.org/analysis/big-winds-dirty-little-secret-rare-earth-minerals/
9. http://www.dailymail.co.uk/home/moslive/article-1350811/In-China-true-cost-Britains-clean-green-wind-power-experiment-Pollution-disastrous-scale.html
10. http://nucleargreen.blogspot.com/2008/02/renewable-and-nuclear-heresies.html
11. Richard S Courtney, Oct. 26, 2006, Lecture and White Paper presented to the Institute of Mining & Mechanical Engineers
12. John Hinderaker, "Why "Green" Energy is Futile, in One Lesson," Thecoachstream.wordpress.com, Feb. 1, 2019
13. Randy Simmons, “What’s the True Cost of Wind Power?” April 11, 2015, Newsweek Magazine
14. Op. cit., Institute of Energy Research
15. http://instituteforenergyresearch.org/press/study-wind-solar-5x-costly-existing-coal-nuclear/

16. cfact.org/2018/05/26/environment-florida-misleads-voters-on-renewable-energ y-costs/?mc_cid=2a94bd2c4c&mc_eid=4e33dab1e3

17. http://www.windturbinesyndrome.com/wind-turbine-syndrome/

18. Ibid.

19. sunvalleysolar.com

20. The Solar Value Cliff: the Diminishing Value of Solar Power", Aug. 2017, Institute for Energy Research

21. "Fast Facts, California ISO, no date

22. http://dailycaller.com/2016/09/22/the-solar-industry-cashes-in-on-government -subsidies/

23. https://chinadigitaltimes.net/2014/06/chinas-solar-panel-production-comes-dirt y-cost/

24. Ariana Eunjung Cha, "Solar Energy Firms Leave Waste Behind in China,"The Washington Post, March 9, 2008

25. Michael Shellenberger, "If Solar panels Are So Clean, Why Do They Produce So Much Toxic Waste?, Forbes Magazine, May 23, 2018

26. https://www.forbes.com/sites/michaelshellenberger/

27. Heller, op. cit.

28. https://hotair.com/jazz-shaw/2022/05/23/its-not-just-electricity-were-heading-f or-shortages-of-diesel-jet-fuel-and-gasoline-n471325

29. Conservation Considerations for Solar Farms, Natural Resources Conservation Service, US Dept. of Agriculture

30. "UPDATED: NEW Vineyard Windfarm Malfunctions. Nantucket Closes Beaches", RI News Today, July 17, 2024.

31. Mike the Writer "Fibreglass from Shredded Wind Turbine Blades Renders Farming Land Useless", Iowa Climate Science Education, June 1, 2024

32. International Energy Agency, 2021

33. Searchinger et al, 2008

34. Peter C. Glover and Michael J. Economides, Energy Climate Wars, Continuum Intl. Publishing Group, 2010, p. 79

35. William O'Keefe, "The Real Reason We Have Ethanol in Our Gas", Fee.org, Nov. 26, 2016

36. http://b3cfuel.com/resources/education/the-problem-with-ethanol

37. David Pimentel, http://hir.harvard.edu/article/?a=1911

38. https://www.statista.com/statistics/281494/us-fuel-ethanol-production/

39. "High Food Price Crisis," www.theoaklandinstitute.org

40. Daniel Strohl, "EPA proposes increasing ethanol content in fuel to 30 percent," https://www. hemmings.com/blog, May 17, 2013

41. https://www.organicconsumers.org/scientific/70-percent-more-energy-required -make-ethanol-actually-ethanol-cornell

42. "Tucker Carlson Tonight", June 10, 2022

43. "Ethanol's Broken Promise: Using Less Corn Ethanol Reduces Greenhouse Gas Emissions", May 29, 2014, www.ewg.org

44. "Did Food Prices Spur the Arab Spring?" https://www.pbs.org/newshour / world/world-july-dec11-food_09-07, Sept. 7, 2011

45. "UN Right to Food Rappeuteur," 2007

46. Georgina Gustin, "Corn-Based Ethanol May Be Worse For the Climate Than Gasoline, a New Study Finds", InsideClimateNews.com, 2/16/21

47. AR. G. Cachat, "Massachusetts: the Hoax of Biomass and Modern Forestry", Aug. 6, 2016, www.the biomassmonitor.orglternative Fuels Data Center, U.S. Dept. of Agriculture

48. Tom Nelson podcast #13, interviewing Tony Heller, Sept. 6, 2022

49. Hayley Smith, "Environmental rules stoke anger as California lets precious stormwater wash out to sea," Los Angeles Times, Jan. 20, 2023

50. Ibid.

51. Ibid.

52. Josh Schlossberg, Ed., thebiomassmonitor.org

53. Josh Schlossberg, Ed., thebiomassmonitor.org

54. Rupert Darwall, "Tesco and the Great Green Scam," https://blogs.spectator.co.u k/2017/05/tesco-great-green-scam/

55. Searchinger, et al., op. cit.

56. Ibid.

Chapter 10

The Energy Transition

> "All the major parties in Europe fear people are at the risk of freezing to death, like what happened in Texas in March of 2021, due to widespread energy outages. But even if human casualties can be avoided, energy poverty becomes an ever more realistic prospect for large parts of the European populace, a risk driven mainly by ideological decisions that were based on very little scientific evidence." — Ralph Schoellhammer, Asst. Professor, Vienna University

Net Zero is defined as a state where the total amount of "greenhouse gas" emissions emitted into the atmosphere is equal to the amount being removed from it. The goal of the IPCC and COP conferences is to get to Net Zero by the year 2050 by using alternative energy sources that don't emit carbon dioxide.

Electric Vehicles (EVs)

As part of the transition from fossil fuels to alternative energy sources, the UN, WEF and other globalist elites have decreed that, as part of the movement away from vehicles powered by internal combustion engines, all vehicles must be run on electricity. This includes not just cars but also trucks, buses and trains. When people complained that gasoline was selling for $5 US per gallon, Senator Stabenow of Michigan responded, "We all need to buy an electric car so we won't have to buy gasoline and be dependent on the whims of the oil companies."

Battery powered cars were popular in the early 1900s. Now they are being promoted as a solution to climate change. But whether or not electric vehicles actually eliminate any carbon dioxide emissions depends on the **source** of the electricity being used to charge their batteries. If that electricity was generated by coal or natural gas — as most power plants still are — there is no benefit to driving an electric vehicle. A study at the University of Michigan concluded that an electric vehicle recharged by electricity from a coal-powered plant generates the same amount of carbon dioxide as a conventional car getting 12.2 km/L (29 miles/gal).[1]Only electricity generated by a "renewable," such as solar or wind power could ever claim to reduce CO_2 emissions.

Even more concerning, a study by the Union of Concerned Scientists revealed that building an electric vehicle generates 15% to 68% more carbon dioxide than the construction of a gasoline or diesel-powered vehicle, due to the intensive process of making its battery. Unfortunately the components of these batteries are not renewable.[2] Every electric car sale represents the energy equivalent of 25 barrels of oil used in making and transporting it.

A recent study[3] found that, in the fourth quarter of 2022 the cost of driving some EVs was higher than similar gasoline-powered cars. The cost of fuel for a mid-priced internal combustion engine (ICE) car was $11.29 US for every 100 miles traveled. This was $0.31 cheaper than the amount paid by drivers in mid-priced EVs being charged mostly at home. This occurred because gasoline prices were getting cheaper and the cost of electricity was increasing at the same time.

Electric vehicles (EVs) are not practical for long distance driving due to range requirements and the absence of recharging stations. Cold weather, which requires using a heater, can substantially reduce an electric vehicle's range. A study by Idaho National Labs revealed that an EV can lose 25% of its rated range when its heater is used in freezing conditions. This would also be true when the car's air conditioner is used. In that study it was also found that fast charging was 35% less efficient at freezing temperatures compared to room temperature.[4]It is no wonder that 90% of the people who purchase an EV also own a fossil fuel powered vehicle that they use for long distance driving.

Recently there was a near disaster in northwest New York. Because the state of New York had mandated that all school bus purchases be electric by the year 2027, the Lake Shore Central School District bought a fleet of them. During an episode of bitterly cold weather in late 2025 the bus drivers were forced to turn off their heaters, which were drastically reducing the driving range of these buses. Otherwise they would have lost power before getting the children home from school. As a result, many of the children were freezing by the time they got home.

In early 2025 only about 2% of vehicles in the United States were battery powered. To reach 100% — fifty times as many battery-powered cars — impossible strain will be put on the grid, which is already being threatened by intermittent wind and solar power. Where do the bureaucrats who are making these decisions think all that extra power will come from? Can the grid really handle 100% of vehicles being EVs? The average annual mileage driven by the 230 million licensed drivers in the US is 13,500 miles. With 33.7 kilowatt hours (kWh) of electrical energy equivalent to 1 gallon of gasoline, this amounts to 4.1 trillion kWh of energy needed, an astronomical figure.

In the summer of 2022 Tesla drivers in Texas were told not to charge their car batteries unless they had no other choice, due to the strain on the electrical grid caused by a heat wave. Summers in Texas have always been hot. 100°F (38° C) summertime highs have been common since the beginning of temperature records. Texas officials should have checked all this out before spending $30 billion to build wind turbines without the permission of the taxpayers who elected them.

At the present time electric cars run on lithium-ion batteries, which have proven to be quite flammable. A fire can start at any moment in one of these batteries. We have seen this happen many times. Manufacturers tried to solve this problem by using smaller cells with firewalls between them. They've tried to produce batteries with components that are less flammable and produce fewer harmful chemicals. Progress has been slow.

It's difficult to extinguish an electric vehicle fire, which needs much more water to put out than a regular car fire. Fires often reignite if the battery hasn't cooled sufficiently. It's important not to keep your electric car in a garage, since many houses have burned down when the electric cars stored in them caught fire. There have been so many of these dangerous fires that hotels in China are starting to ban EVs from underground parking lots.

"It's a serious issue that needs to be solved," lamented Dave Sargent, head of the automotive practice at research firm J.D. Power. "It's something the industry will have to get a grip on if EVs are going to be the future."[5] Any lithium-ion battery, even a cell phone battery, can catch fire. Manufacturers have recalled some of their car models and are working hard to find a solution to this problem.

Instead of taking five minutes to refuel as with an internal combustion engine, it could take two to eight hours to charge an electric vehicle battery, depending on the charger. Battery-powered cars average 3-4 miles per kWh of electricity. If you charge your car at home (as most people do), and drive 1,000 miles per month, at the price of 14.47 cents per kWh, you would pay $36.29 extra to the electric company. In California, with the average price of a kilowatt hour at 22.81 cents, your cost would be $57.03 per month with the 8-hour charger. "A level two EVSE requires special charging equipment and a separate, more powerful electrical circuit. You would also have to pay installation costs. On average, a level two EVSE system costs $400 to $6,500 for equipment, and $600 to $12,700 for installation."[6]A study by the Anderson Economic Group concluded that it is cheaper to operate a midsize fossil fuel powered car than a comparable EV.[7]

One of the worst aspects of battery-powered vehicle ownership is that these vehicles can be controlled from an external source. Governments are requiring all car manufacturers to place a "kill switch" in each vehicle by 2026; many already have these. Remotely activating this switch will render the car inoperable. This means that the owner of the car doesn't have ultimate control of it. Manufacturers are already charging consumers extra per month for heated seats, an option that is controlled from a remote location. Cars can also be set up so that their speed doesn't exceed 65 mph or some other speed.

Electric car manufacturers are raising vehicle prices due to the skyrocketing costs of some of the key materials needed for their batteries. The International Energy Agency reported that although battery costs had been declining for a decade, they are now expected to increase by 14%. One manufacturer blames the increased cost on the projected demand for minerals rising another 20% over the next four years. The five principal minerals used in manufacturing EVs are cobalt, lithium, graphite, nickel and rare earth metals, exceeding all other uses of these metals combined.

Governments are sinking tremendous subsidies into the purchase of these cars. An EV purchaser in the United States receives a $7,500 discount paid by taxes. Most battery-powered cars are bought by the richest 20% of the population. This is welfare for the rich at the expense of middle-income and poor people.

In 2025 the average price of a battery-powered car was US $57,245, compared to about $45,000 for a car powered by an internal combustion engine. [8] With 60% of Americans living from paycheck to paycheck, most cannot afford to pay almost $60,000 for a vehicle, even though the US government will reimburse EV buyers $7,500, which is paid for by the taxes of the people who can't afford battery-powered vehicles. The federal government appropriates tens of billions of dollars of subsidies to battery producers and for the construction of millions of new electric vehicle charging stations.[9]

The Environmental Protection Agency previously released new tailpipe emission rules that were so strict that cars with internal combustion engines (ICE) wouldn't be able to comply. This is how the Biden administration intended to force Americans to buy electric cars. Under the Trump administration these requirements were relaxed. If a government promoting climate change mitigation assumes power at the end of the Trump administration in 2029, the US will revert to the repressive policies of the Biden administration.

Norway has perhaps progressed further along the energy transition route than any other country. This is because 88% of its electricity comes from hydroelectric generating plants: 1,791 of them. Nearly 96% of all new cars in Norway are battery powered. Their owners receive all kinds of benefits from the government, including a 25% value added tax exemption, a 70% discount on toll roads and a 50% discount on ferry fares.

France comes in second, with over 70% of its electricity being generated by nuclear energy, much higher than the world average of 10%. Most climate change activists want no part of nuclear energy. They have generally lobbied against it, due to safety concerns. They don't seem to understand that nuclear energy is by far the safest and the least environmentally invasive of all methods of electricity generation.

The Grid

A vast network of cables transporting electrical power from generators to electricity users, the grid will have to be greatly expanded — nearly doubled — in order to make energy from "renewables" possible. This essential part of the system seems to have been forgotten by most policymakers. In many cases work on its expansion hasn't even started. To hook up all of these new power sources to where they can be used, the grid must be lengthened by about 150 million km (92 million miles), or the distance from Earth to the sun.

A nuclear power plant producing one gigawatt of power would essentially require one grid hookup. But 800 giant wind turbine generators, producing about the same amount of power as that one nuclear plant, would need 800 grid connections, greatly multiplying the amount of cable needed.

This project will require a massive investment of about $21 trillion US. The work will have to be finished within the next ten years if proponents are really serious about reaching Net Zero by 2050. Also, the current grid needs to be upgraded, since 30% of it is over 50 years old. Doubling the size of the grid will require about 427 million tons of copper, which is in short supply.

businessinsider.com

According to Mark Mills, CEO of the National Center for Climate Analytics, the seven-fold increase in demand for minerals required for the widespread adoption of not just electric vehicles but lithium ion batteries used to back up solar and wind installations will be the greatest ever in all of human history. There is already a shortage of copper, cadmium, nickel, cobalt, zinc, aluminum and rare earth metals. Since copper is also used for electrical cables, a serious shortage of that metal is expected in a year or two. An electric vehicle needs 136-182 kg (300-400 lbs) more copper than a vehicle with an internal combustion engine.

It is estimated that only about one tenth of the needed minerals to achieve Net Zero are currently available.

It all starts with mines. "A mine is a giant hole in a fragile ecosystem." [10]There are serious shortages of lithium, cobalt, graphite and nickel. It's not that the earth lacks these resources, but mines must first be established before they can be extracted. It takes about sixteen years to establish a new mine.[11]We will have run out of some of these crucial minerals before new mines can be established. Each new mine requires billions of dollars of capital investment, which is not possible without government subsidies. Therefore, 2030 will come and go before we ever have the energy transition that so many people are expecting. In North America and Europe very few new mines are being established. "We don't have the minerals [for Net Zero], and we can't afford it," said Mark Mills.[12]

Over half of the world's cobalt currently comes from the Democratic Republic of the Congo, one of the poorest countries on Earth. China is directing many of these mining operations in Africa. Recently it was revealed that **thousands** of children in the Congo, some as young as four years of age, have been put to work in the arduous, dangerous and dirty process of mining for cobalt under horrendous conditions. Anyone who purchases a product with such a battery knowing that s/he is contributing to this atrocious practice is complicit in the abuse of these children. According to Bloomberg, "For consumers seeking to lower their carbon footprints, the environmental and social costs of electric vehicles may be greater than they realize." Many of these materials are produced in countries such as China and Iran that are known for serious, widespread human rights violations.

Children Mining Cobalt in the Congo

Mining and processing the materials needed for EV batteries is an environmental disaster. "The production of *just one* EV battery requires 11.4 kg (25 lbs) of lithium, 27 kg (60 lbs) of nickel, 20 kg (44 lbs) of manganese, 13.6 kg (30 lbs) of cobalt, 91 kg (200 lbs) of copper, and 182 kg (400 lbs) of aluminum, steel and plastic. 11,363 kg (25,000 lbs) of brine is required to extract the lithium, 13,600 kg (30,000 lbs) of ore for the cobalt, 2,270 kg (5,000 lbs) of ore for the nickel, and 11,360 kg (25,000 lbs) of ore for the copper. Huge diesel machines that burn hydrocarbons are being used to extract this earth.

These electric car batteries are extremely heavy. A car with an internal combustion engine has a gas tank with 36 kg (80 lbs) of fuel in it. An electric vehicle has a 455-kg (1,000-lb) battery. All told, about half a million pounds (227,000 kilograms) of the earth's crust must be dug up for just one battery."[13] A heavy-duty electric truck needs a battery weighing up to 7,270 kg (16,000 lbs) — 16 times as heavy as a Tesla car battery. Eight tons of ore must be dug up to provide the minerals for such a battery. [14]

As the wind industry has grown, so has the demand for neodymium, which could increase by as much as 700 percent over the next 25 years, and for dysprosium, which could increase by 2,600 percent. An article in *The Daily Mail* stated that every wind turbine contributes to "a vast man-made lake of poison in northern China."[15]

A huge amount of carbon dioxide is generated in manufacturing, transporting and installing wind turbines. The foundation of **just one industrial wind turbine** contains over 500 cubic meters of concrete, emitting 624,000 kg (1,372,800 lbs) of carbon dioxide in its production. This doesn't include the CO_2 emitted in the transportation of this concrete to the job site. This fact reveals the hypocrisy of those who claim that wind turbines are good for the environment. The scrapping of wind turbines that have worn out after their 10–15 years of useful life are not included in any CO_2 cost-saving calculations, making emissions claims a farce. Many older wind farms that have outlived their usefulness have simply been abandoned, because properly disposing of them would be prohibitively expensive.

China controls two thirds of all the minerals needed for energy. Currently the United States is dependent on China for 100% of rare earth metals. A couple of decades ago the US made a decision to discontinue processing these rare earth minerals, and China took over their extraction and refining. The Trump administration is trying to remedy that situation.

"We've spent trillions of dollars to avoid oil and gas, and global demand despite all that spending and all that stuff being built in the millions of EVs going to market, oil demand is higher than it was before the [COVID] lockdowns," Mark Mills pointed out. [16]

Sixty percent of copper and seventy percent of lithium is refined in China. Also, one hundred percent of graphite is refined in China. The United States invented that process and taught it to the Chinese. 90% of all solar panels are manufactured in China with US help. This process is incredibly energy intensive and accomplished with coal-fired operations. China has the needed energy resources, because they've been building 100 coal-powered generators every year. In contrast, the United States has shut down 300 coal-fired plants. At this time China burns more coal than all other nations put together: 58% of the world's total. They've announced that they will not start curbing their CO2 emissions until 2050.

In Indonesia, where nickel for EV batteries is produced, "polluted air and water are causing respiratory problems, sickness, and eye injuries and destroying forests and fisheries. The rush to expand production has pushed some local communities and infrastructure to the brink of collapse."[17]It's a big, dirty process using lots of chemicals — really shocking.

Lithium Mine - courtesy of survivethenews.com

Lithium mining is one of the most ecologically destructive on the planet. At just one lithium mine 35-40 Caterpillar haul trucks use up almost 18 million gallons of diesel — a hydrocarbon fuel — per year. Lithium consumes vast quantities of water that is contaminated in the process and unable to be reused.

The production costs of nickel and cobalt (also used in smart phones, laptop computers and some power tools) have multiplied by several times during the last seven years. Because Russia is a major nickel exporter, its invasion of Ukraine is interfering with the production of nickel.

Battery-powered vehicles also use much more aluminum than conventional vehicles. Ford's F-150 battery-powered truck contains 682 lbs of aluminum, mostly mined and refined in Brazil. There is a class action lawsuit going on there against Hydro, a Norwegian company. "The alumina made here is produced at the cost of a lot of misery," proclaimed Ismael Mores, a lawyer working on this lawsuit. Here bauxite (aluminum ore) is mined and converted to alumina and a toxic residue which leaches into rivers and creeks.

In recent Congressional testimony Mark Mills reported that before an EV is sold, it has already consumed the energy equivalent of 20-30 barrels (840-1260 US gallons) of petroleum. He noted that two dozen minerals are needed in the manufacture of a battery-powered vehicle, which requires four times as many minerals as a traditional car.[18]Copper, which is needed for many things including the electric grid, is in very short supply right now and will have to be increased by two and one half times. Although ore containing copper is still available, it is a lower grade ore, not as rich in that mineral as previous amounts that have already been used. This will require more ore to get the same amount of copper. To extract copper in Chile, arsenic-laden soil is blown up. Arsenic poisons crops, and percolates into lakes and streams, poisoning fish, livestock and people. There is a high incidence of cancer in the area surrounding these mines.

In order to get to Net Zero the world will have to increase the number of mines tenfold. This means that an estimated 400 new mines will be needed.

Lithium-ion batteries are usually replaced every seven years for smaller vehicles and every three to four years for larger vehicles, such as vans and buses. Over a period of time, the number of miles a car can be driven between battery charges declines. The car owner must then recharge the battery more and more frequently. The guarantee offered for such a battery is honored only if it wasn't damaged by an incidental problem, such as a collision or a stone flying up and damaging one of the 3,000 or so cells making up the battery. There have been cases where a relatively new battery had to be replaced because battery cells had been accidentally damaged.

Because batteries are full of toxic chemicals, they can't just be placed into landfills. Like solar panels, they require recycling, which is an intensive process. Recycling these batteries is difficult and complicated due to the sophisticated chemical procedures involved. The cost of recycling really needs to be included in the price of an electric vehicle. If batteries are not properly recycled, the heavy metals contained in these will contaminate water and soil. Extracting lithium from a used battery is five times as expensive as the cost of mining it. So there is not much incentive to recycle batteries.

These problems seem almost too unbelievable to be true, but in the writing of this book, the truth has been adhered to as much as possible. If you doubt any of these figures, it is our hope that you will investigate further and let us know if you find any of these conclusions to be mistaken.

Inventors are working feverishly to correct these problems and develop electricity storage systems that can store more power for longer periods of time. But this may take many years if indeed these problems can ever be solved. How much time will pass before this happens is unknown. Those in power should have had **all** of these problems worked out **before** putting the world's people in a no-win situation. What they have actually done instead is like jumping out of a plane with no parachute.

The Real Cost of Net Zero

A video by Business Insider [19]claims that between 2010 and 2024 the cost of solar panels decreased 90% and wind installations by 60% and that wind installations are cheaper to build than new natural gas plants, even without tax credits. [20]While it is true that the price of solar panels has come down, it is absolutely untrue that commercial installations are cheaper than conventional power. According to low-end estimates, in 2025 a wind turbine rated at 2 MW costs about two million dollars (US) to build. 500 of these 2-MW wind turbines would be rated at one gigawatt (1,000 MW) and cost about $1 billion US. Because the efficiency rating for such a wind turbine is only about 22% of its nameplate rating, the actual power delivered is only about 220 megawatts (MW).

A combined-cycle natural gas generator rated at 1 GW (billion watts) also costs about $1 billion to build and, with an efficiency of 60% provides 600 MW of power. Therefore, an equivalent amount of power from wind energy costs 2.73 times as much to build as a natural gas power plant. That doesn't include the cost of the land. Each turbine requires about 7.29 hectares (18 acres) of land. 500 turbines which, together generate about 220 megawatts, would require an area of 3,645 hectares (9,000 acres). At an average cost of $3,000/acre the cost of the land alone would total $27 million US. But perhaps the land can be leased.

But that's just part of the story. Coal, natural gas and nuclear electric generators have a lifetime of fifty or more years, whereas the life expectancy of a properly maintained wind turbine is only about 15-20 years. They are expensive to maintain, costing about $50,000/year per turbine —$25 million/year X 20 years = $500 million, half the cost of the installation. Of course natural gas plants also need maintenance, but not as much.

To prevent electricity blackouts and brownouts from occurring when the sun isn't shining and the wind isn't blowing at just the right speed, wind and solar installations must be backed up by gas plants (called peaker plants) which must be rapidly activated during these times and which add dramatically to the cost.

This is what people don't realize, because they haven't been given accurate information. The sun and wind are free. But the question is: how much material is required to build all of the machines necessary to convert it to a usable form to extract (really, to gather) energy from sunlight and wind? **Embodied energy** is the energy required to produce an item. Batteries store energy, but how much energy does it take to make a battery? It turns out that it takes the energy equivalent of **300** barrels of oil to build a battery large enough to store just **one** barrel of oil equivalent.

This video blames a number of factors, including high demand for energy, the aging grid, bureaucratic regulations, and even **climate change itself (!)** for higher electricity costs. They also contend, "The states with the largest increase in solar and wind between 2019 and 2024 appear, if anything, to have experienced price decreases over the same period."

How could the electricity bills from "renewable" power sources be lower, when a map of the United States clearly shows that electricity costs have increased the most in the states mandating the biggest increases in "renewable" generation? "Each of the top five most expensive states for electricity have mandates requiring 100% of their power to come from renewable or carbon-free sources."[21] A study by the Progressive Policy Institute determined that New York state has experienced some of the fastest increases in electricity rates in the nation, rising 36% from 2019 to 2024, three times faster than the national average.

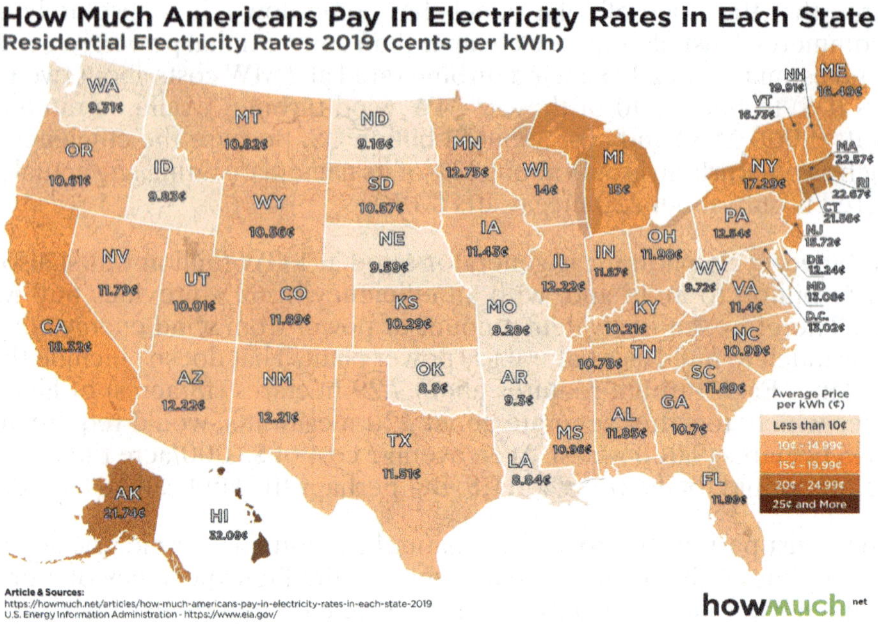

And then we can compare the mineral requirements of different energy-producing systems for the same amount of electricity. The basic materials required to produce **one terawatt hour** (TWh) of electricity, are: 16,000 tons for solar, 10,000 tons for wind, 5,000 tons for geothermal and just 2,000 tons for natural gas. To produce the same amount of electricity as a 100-MW natural gas power plant the size of a tractor trailer, an array of solar voltaic panels requires **hundreds of acres and fifteen times as much tonnage of materials** to build. Solar panels wear out in less than twenty years, less than half the lifespan of a coal, gas or nuclear plant.

To supply half of the world's electricity with wind turbines two billion tons of coal would have to be burned just to produce the concrete and steel used to build them. Steel cannot be made without using coal. There is no way this can be done with solar, wind or nuclear energy.

To fulfill their emission goals Germany and Great Britain have drastically deindustrialized. Britain has suffered a severe economic decline in the last few decades due to the drastic decline of its steel industry. Having shut down their last coal power plant, Britain is no longer able to make steel.

In the North Sea off the British coast, Hornsea3, still under construction, is the largest wind power installation in the world. It is projected to provide 2.9 GW of power to service three million homes at an estimated cost of 8.5 billion pounds. This would be correct if it were running at 100% capacity. But at 30% capacity it would provide only 10 kWh per home each day. The average home uses 90 kWh hours per day. [22] To achieve the goal of being all electric by 2050, Britain would need 300 GW of electric power, taking 100 Hornsea3s, and costing 1.8 trillion pounds, not including backup storage. [23]

The amount of energy storage required depends on the maximum length of time that wind isn't blowing at the correct speeds. The longest wind drought they've experienced is 21 days at just 10% of peak wind energy. In a case like this ten days of full backup storage would be required. The cost of a battery bank large enough to supply this is 10 trillion pounds. This is why the operators of these wind farms prefer backup from gas power plants over battery storage.

In the winter in the UK solar panels are operating at just 3.6% of installed capacity. In the summer it is 23%. But these are being sold to the public as 100% of installed capacity — 24 hours a day and seven days a week. This is pure disinformation (lying).

Britain has the highest energy prices in Europe. David Turver, an independent energy analyst, estimates that Britain's cost of going to net zero would be at least nine trillion pounds.

California has the highest energy prices in the US. "Free" energy is not free.

On February 24, 2022, Russia invaded Ukraine. Despite all the hype from the Paris Accord about converting from fossil fuels to renewables, Europe is still heavily dependent on Russia for its energy needs. Europe produces 3.6 million barrels of oil per day but uses 15 million barrels per day. Russia supplies 20% of Europe's oil, 40% of its natural gas and 20% of its coal.

The Centre for European Economic Research reported that just between 2023 and 2024 Germany lost nearly 200,000 companies, especially energy intensive industries. [24]One of these is the Volkswagon corporation. A German steel company moved from Germany to Texas. You can't make concrete or steel if you don't know how much power you have. "Germany's energy transition has driven prices to unsustainable levels, pushing companies abroad. Ending the energy transition and returning to affordable energy is vital for industrial survival." [25]Germany shut down its last nuclear plant in 2022. Since 2002 Germany has spent $744.7 billion US on subsidies for "renewables".

If Germany stays on its present course, by the year 2045 it will have spent $6.4 trillion US on climate change mitigation. This development has led to the rise of parties such as the right-wing Alternative for Germany (AfD). Fortunately, many people in Germany are waking up and realizing that they must change course in order to survive.

Ralph Schoellhammer observed, "They've been trying the energy experiment in Europe for thirty years." He contended that the experiment should have first been tried on a small scale instead of with the entire world.[26]What's been happening in Europe and Australia should be a wake up call for countries that are not as deep into the "transition".

The Disastrous Descent into Net Zero

Another case in point is that of Washington and Oregon. In 2016 Governor Inslee of Washington commissioned a report from Evolved Energy Research, which concluded that converting power sources to "renewables" — an 80% reduction in greenhouse gases by 2050 — would save the average household $7 per month over what they were then paying for electricity. Audubon Washington chimed in, "A transition to 100% clean energy is achievable, affordable and reliable."

Another organization, Energy and Environmental Economics, Inc., reached a different conclusion, stating, "Removing the final 1% of carbon requires an additional $100 billion to $170 billion of investment" (in 2016 dollars). The Washington Dept. of Commerce decided that it was "premature to assess impacts of 2030 and 2045 standards." So in 2019 the Washington state legislature passed the "Clean Energy Transformation Act" (SB 5116).

Not to be outdone by its neighboring state, the Oregon legislature passed a bill (#2021) calling for Oregon to be 100% carbon free by 2040. They used no reports outlining the costs of such an endeavor to guide this decision.

Both of these states have adopted California's Clean Car Rules, requiring 100% of all cars and light trucks to be battery powered. This in itself would drastically increase the demand for electricity.

Concerned about the real costs of getting to Net Zero in these states, Jonathan Lesser and Mitchell Rolling conducted a study that was published in 2024. [27]This study concluded that by 2050 peak demand would likely double over the 2023 peak demand due to the full electrification of vehicles and conversion of all residential heating to electricity. Building the needed "renewables" would require a shocking 25% of the land area of eastern Washington and Oregon, not including the land required for transmission lines.

This report included the fact that peak demand for electricity occurs in the early evening hours when solar radiation diminishes and wind speed often subsides. A huge amount of large scale battery storage capacity would be required to prevent blackouts during these time periods. They determined the average useful life of battery storage to be fifteen years. This report concluded that projected costs of these project will amount to $682.8 billion by 2050 plus installation costs. This conservative estimate including finance costs and land totals $1.011 trillion.

There are 4.85 million households in Washington and Oregon. If this report is correct, the cost per household will exceed $200,000. An average electric bill for residents will increase by 450%: from $110 per month to over $700 per month, assuming a modest 2% annual inflation. The rate for commercial customers will increase from $600/month to $3,800/month. The elimination of all greenhouse gas emissions by these states in 2040 will decrease the earth's average temperature by just 0.003^0 C. This figure is 170 times less than the accuracy of the best outdoor thermometers, meaning that the temperature difference will be indistinguishable. Yet such an expense would wreck the economies of these two states.

Carbon Capture and Sequestration

Carbon capture and sequestration (CCS) is a process that captures carbon dioxide gas from power plants before it enters the atmosphere and transports it to a storage location, typically deep underground, or in depleted oil and gas reservoirs, for long-term isolation.

CCS is an expensive process requiring a great deal of energy, up to 25% of the power generated by a coal-fired power plant. [28]Currently the methods for doing this are not very efficient. Improving capture efficiency is crucial to the development of this process. The operating costs of CCS are very high, ranging from US $40-100 per ton of gas captured. The physical footprint of these facilities is very large, requiring a significant land area that impacts the local ecosystems and communities and adds significantly to the cost.

The pipelines used to transport carbon dioxide underground are expensive to build and maintain. This presents a significant challenge, because corrosion can occur. Before CO2 can be transported, it must be liquified, requiring substantial energy. Longer transport distances increase the costs of compression, liquefaction and pipeline structural integrity. Injecting CO2 into underground storage can contaminate local groundwater resources, posing a potential threat to human health. There is a limited number of storage sites available. For these reasons, Carbon Capture and Storage is not considered a viable solution to the climate change "problem".

According to Steve Goreham, carbon capture technology isn't ready despite the demands by the Biden administration that it be used with new power plants.[29]In order to reach net zero by 2050, a billion tons of carbon dioxide **per year** would have to be trapped underground.

The Geoengineering Threat

Those pushing the climate change narrative are beginning to realize that the measures taken to cool the earth by reducing carbon dioxide are not working as well as they had hoped. So now they've decided to take more extreme measures.

One plan, dubbed Solar Radiation Management (SRM), also called Solar Geoengineering, would inject millions of tons of tiny sulfur dioxide (SO_2) particles into the stratosphere at an altitude of 16-24 km (10-15 miles) to reflect sunlight in an effort to cool the planet. A 2018 Harvard University study estimated the cost of doing this to be around US $2.25 billion per year over a 15-year period.

A California project, entitled "Make Sunsets", released sulfur and other harmful chemicals into the atmosphere using high-altitude weather balloons. When they tried to move this project to Baja California in Mexico, the project managers were met with criticism and outrage, and the Mexican government put a stop to it.

An Israeli company called Stardust Solutions is testing a plan to shoot reflective particles into the sky at a high altitude to try to reduce solar radiation. But some fear that the ozone (O_3) layer in the stratosphere could be harmed by reactions with these particles. Because ozone absorbs ultraviolet radiation from the sun, more UV radiation would thus reach the earth's surface.

Other geoengineering projects are in the planning stages or already being implemented. A project at Southern Cross University in Australia involves blasting a brine mixture into the sky to convert wispy clouds into larger, brighter clouds that would supposedly reflect more sunlight back into space. This endeavor is being funded by the government, various conservation groups and universities. Their hypothesis is that "brightening clouds" would increase the albedo effect — with light-colored surfaces reflecting light — preventing it from reaching the ocean surface. The brightness of a cloud depends on the size of each water droplet in it. The smaller the droplets the more surface area there would be to scatter light rays.

To achieve this, tiny aerosols would be sprayed on the ocean surface. These would provide condensation nuclei to form smaller water droplets than would form naturally if the water evaporating from the ocean were allowed to rise by convection until it reaches an altitude where it is cool enough to condense. A large fleet of ships would be required to carry the machines needed to do this. Building the required ships would cost upwards of $5 billion US.

Some worry that reducing the droplet size would affect the period of time that these clouds exist, as well as the amount of water they can hold. The risk here is that weather patterns could be affected. Since scientists simply don't know enough about how climate works, this experiment could have negative, unintended consequences. Playing with the earth's thermostat could change rainfall patterns and shift monsoons, with potentially devastating consequences for agriculture. The resulting unintended consequences in climate could vary between regions, with some areas benefiting and others being harmed. This could create conflict between nations.

At Martha's Vineyard, in the United States, the Woods Hole Oceanographic Institute is planning to dump thousands of gallons of sodium hydroxide, a strong alkali mixed with dye into the ocean. They claim that this will create a carbon sink, to draw carbon dioxide from the atmosphere into the ocean from the atmosphere — a project known as Ocean Alkalinity Enhancement (OAE). Other scientists, as well as fishermen are concerned about the effects this action would have on marine life. Woods Hole scientists recently applied for a permit from the EPA to do this. [30]

One of the best studied methods of geoengineering is ocean fertilization. This involves increasing the growth of phytoplankton, tiny algae floating on the ocean surface, that conduct photosynthesis. About 50% of all photosynthesis on Earth is done by phytoplankton. These can be fertilized to increase photosynthesis. These tiny plants would use up more carbon dioxide in the process. Iron is the main fertilizer being considered for this project. However, this could create an overgrowth of phytoplankton, that would deplete oxygen at the ocean surface, harming marine animals which depend on it. A sudden growth of phytoplankton could thus destabilize marine ecosystems, interfering with the food chain.

A study by NOAA released on Feb. 28, 2024, calls for removing water vapor from the stratosphere. In other words, this would dehydrate the atmosphere. After all, the main greenhouse gas is water vapor, which has over seventy times the effect on the earth's atmosphere as carbon dioxide. The authors propose that this could be done by seeding tiny "ice nuclei" into the atmosphere with high altitude planes. They believe that this ice would cause the water vapor in the stratosphere to crystallize into ice, which would then fall to lower altitudes.

The study's abstract explains, "The mechanism is the targeted injection of ice-nucleating particles (INP) in air supersaturated with respect to ice at high altitudes in the tropical entryway to the stratosphere. Ice formation in this region is a critical control of stratospheric water vapor (WV). Recent airborne in situ data indicate that targeting only a small fraction of air parcels in the region would be sufficient to achieve substantial removal of water." The lead author of this study, Joshua Schwartz, admits that currently NOAA doesn't have the technological ability to accomplish this feat.

Just like the study of climate, the technology of geoengineering is still in its infancy. Experiments done with climate models can't tell us just how these measures would affect the real world. The prospect of creating a "solution" that is orders of magnitude more dangerous than the "problem" is so scary that even some climate change believers are warning against it. "The ethical questions associated with climate manipulation loom so large that some forms of geoengineering are simply unacceptable," wrote Kevin Trenberth, in the journal, *Physics Today.* Kevin Trenberth has been one of the main proponents of the climate change narrative.

There are signs that the Net Zero agenda is losing its grip on the world. A Net Zero Alliance is any global group of financial institutions, asset holders or other stakeholders committed to achieving a net-zero economy by the year 2050. Since December of 2024, in an apparent reaction to criticism of ESG, Citibank, Goldman Sachs, JP Morgan, Morgan Stanley and Wells Fargo left the Net Zero banking alliance. Climate Action 100+ is the largest and most aggressive investor initiative on climate change. In May, 2024, this organization lost three of the world's six largest management funds.

Recently Bill Gates, one of the most outspoken climate change advocates, radically reversed his position on the issue,[31] stating that climate change is not a crisis after all and that we should instead focus on human welfare. One reason for Gates' change of heart could be his interest in artificial intelligence (AI), which has recently emerged onto the scene and which requires a colossal amount of electric power. There is a mad scramble for AI data centers going on right now. 1.5 million AI server units would need at least 85 terawatt hours of electricity per year, an unbelievable amount of power. This could never be possible with wind and solar, and Bill Gates wants AI. He is working on restarting the Three Mile Island nuclear facility in Pennsylvania.

On Nov. 19, 2023, Shane Shifflet of the Wall St. Journal noted, "Wall St. rushed to embrace sustainable investing just a few years ago. Now it is quietly closing funds or scrubbing their names after disappointing returns that have investors cashing out billions."

Since 2018 some investors have been warning that the wind and solar industries would ultimately collapse due to their reliance on subsidies. *Venture Capitalist* predicts that "An Enron-style collapse is in store for renewables."

For all of the above reasons it is extremely unlikely — and some say impossible — that the energy transition to "renewables" will ever take place. "We're engaging in a dangerous, economically and geopolitically dangerous fiction around a narrative that is as close to a modern fairy tale — the emperor has no clothes. There is no energy transition," Mark Mills, founder and CEO of the National Center for Energy Analytics declared.

---

1. Michael Sivak & Brandon Schoettle, "Relative costs of driving electric and Gasoline Vehicles in the Individual U.S. States", Feb. 2018.
2. Editorial, "Investor's Business Daily", 11/7/17
3. by Anderson Economic Group
4. https://www.cnet.com/roadshow/news/electric-vehicle-range-shopping-tips/
5. https://www.forbes.com/wheels/news/battery-car-fires/
6. https://www.investopedia.com/how-much-does-it-cost-to-charge-an-electric-car-5186285

7. Jamie L. LaReau, "Which is more expensive: Charging an Electric Car or Fueling a Car with Gas?" Detroit Free Press, 10/21/21

8. Renee Valdes, "How Much Are Electric Cars?", Kelley Blue Book, 9/17/25

9. Editorial, Institute for Energy Research, 4/14/23

10. Mark Mills, op. cit.

11. Mark Mills, op. cit.

12. Mark Mills, Ibid.

13. Ronald Stein, "Is it Unethical to Buy a Vehicle Powered by a Lithium Battery?" 6/13/22 "Issues in Insights "

14. Ronald Stein and John Shanahan, "The 'Green' Scam of the Century: How 'Renewables' Increase Fossil Fuel Demands," *Climate Change Dispatch,* Oct. 23, 2024.

15. https://www.instituteforenergyresearch.org/renewable/wind/big-winds-dirty-littl e-secret-rare-earth-minerals/

16. Mark Mills, op cit.

17. Ibid.

18. www.instituteforenergyresearch.org > renewable > magic-renewables-mark-mills-revisited

19. "You're Being Lied to About Electricity Costs | Truth Complex", Business Insider https://www.youtube.com/watch?v=pk0Zm_xpk3c&t=237s

20. "You're Being Lied to About Electricity Costs | Truth Complex", Dec. 20, 2025

21. Thomas J. Pyle, Kenny Stein, Alexander Stevens, "Blue States, High Rates," Institute for Energy Research (IER), Dec. 10, 2025.

22. Ralph Ellis, "Net Zero Nonsense", Tom Nelson Podcast #289, March 28, 2025.

23. Ibid.

24. J.C. Kofner, "A New Industrial Policy for Germany: Reshoring Through Targeted Incentives," MIWI Institute, May 26, 2025.

25. Ibid.

26. "Is Europe in Decline? A Conversation With Ralph Schoellhammer in Miskolc, mcc.hu, Jan. 29, 2025.

27. Jonathan Lesser and Mitchell Rolling, "The Crippling Costs of Electrification and Net Zero Energy Policies in the Pacific Northwest," The Discovery Institute, Sept. 2024.

28. Global CCS Institute, 2021

29. Steve Goreham, *Green Breakdown: the Coming Renewable Energy Failure,* Ste veGoreham.com

30. Victor Tangermann, "Scientists Seeking Permit to Dump Chemical Into Ocean in Experiment to Stop Climate Change", Futurism.com, July 15, 2024.

31. Amy Harder, "How Bill Gates is Reframing the Climate Change Debate", *Axios,* Oct. 30, 2025

Chapter 11

The Vision of Our Energy Future

"It disturbs me hearing people like Al Gore say that he is worried about the future of his grandchildren. Complete madness." —Dr. Richard Tol, UN IPCC Lead Author

Increasing Demands for Energy

Despite the efforts of climate change activists to control citizens by lowering living standards and pushing for outrageous actions such as moving to 15-minute cities, getting rid of our pets, ending meat consumption and controlling every aspect of our lives, the demand for energy isn't just rising — it's accelerating by leaps and bounds. A great deal of this increase is due to the emergence of artificial intelligence (AI). There seems to be no stopping it. AI data centers are extremely energy hungry. Hundreds of data centers are being built in the US. Each requires hundreds of megawatts of power. AI will soon be using about 25% of **all** electricity in the US. Each data center can easily consume as much energy as a city.

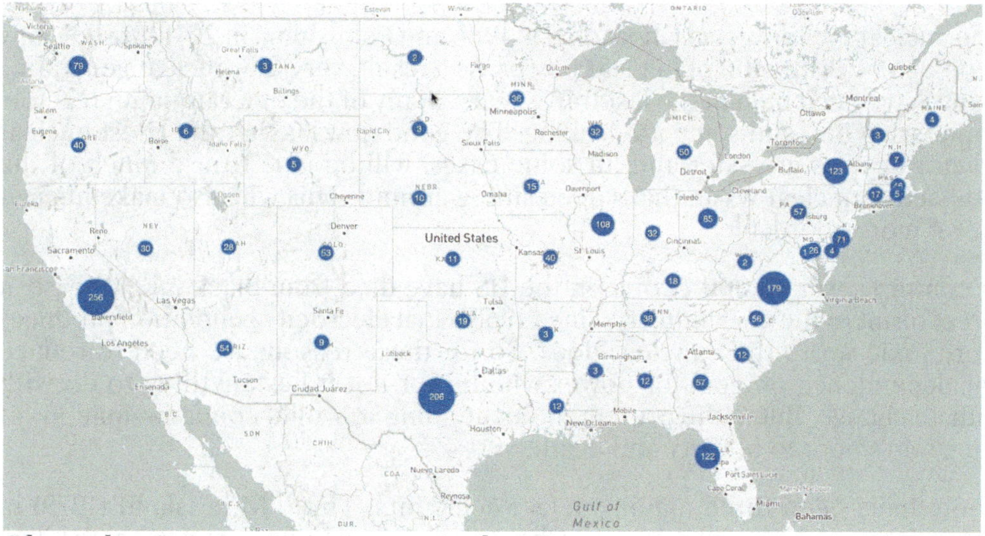

Planned or Existing Data Centers in the US

This is a reality that can't be ignored, and it changes the character of the climate change issue. Currently and in the near future the electricity supply for data centers will come from natural gas. But ultimately it will have to be supplied by nuclear power.

Crypto mining to produce cryptocurrency is another gargantuan energy user. It uses up to 2.3% of all energy in the United States. These "mining facilities" that draw power from the grid have already threatened to cause brownouts and blackouts during periods when demand was high. In Texas this practice raised electricity costs by $1.8 billion. In Nebraska and Kentucky utilities were required to buy expensive grid upgrades to accomodate crypto miners at everyone else's expense.

If our federal and state policymakers believed that there was a climate crisis, this practice would never be permitted. Have the lobbyists and other rich people who profit from cryptocurrency been twisting arms? Who is behind this?

A novel idea being proposed to provide power for AI data centers is to put them in orbit — a special type of "sun-synchronous orbit" known as dawn-dusk SSO — powered by an array of photovoltaic panels. In this setup the solar panels would be receiving sunlight 95-100 percent of the time. This setup would avoid night, seasons, weather and atmospheric filtering. Being in a vacuum, they would require no fans or water for cooling. This project would require a massive installation of gigawatt-scale data centers within one or two decades, using many satellites. Fortunately, SpaceX starships are reusable. Communication between satellites and with the ground would be accomplished using lasers with low latency. Signals would reach their destinations within just a fraction of a second.

Nuclear Energy: Our Best Option

At the present time nuclear energy is producing about 19% of the electricity in the US. The United States currently has 54 commercial nuclear power plants, the average age of which is about 42 years.[1] Many of the older nuclear generators are being decommissioned and not replaced because of a lack of understanding of the dangers of nuclear energy, which has also been aggravated by the legacy media.

The nuclear meltdowns at Chernobyl in 1986 and Fukushima in 2011 caused many people to be distrustful of nuclear power. As a result, very few nuclear generating plants have been built in the last thirty years. Many of the older uranium reactors are wearing out, and some are leaking. The good news is that the latest nuclear reactor designs are safer and, in some cases, will be able to use much of the radioactive nuclear waste that's been sitting around. This will also make nuclear power more affordable.

In contrast, over 30,000 people in the US have died from black lung disease, a direct result of the coal industry, since 1968. Coal electricity generation produces more toxic waste than nuclear does. So for these reasons we need to reduce our dependence on coal for power whether or not it has anything to do with climate change. But at the present moment doing so would condemn millions of the world's poor to poverty and death.

There have been three major nuclear accidents: Three Mile Island (1979) in Pennsylvania, which killed no one, Chernobyl (1986) in the Ukraine, which killed about 50 people, and Fukushima (2011) in Japan, which killed no one. Considering the amount of electricity generated worldwide by nuclear reactors, there have been far fewer casualties from nuclear power than with any other source of energy. No one was killed or injured in the partial core meltdown at Three Mile Island in 1979, and it leaked no appreciable radiation. This plant's containment system worked.

Although the Fukushima accident was much more serious, no one died from radiation exposure. This accident resulted from a tsunami caused by a tremendous undersea earthquake registering 9.0 on the Richter scale. This was one of the most violent quakes in recorded history. The ensuing series of waves penetrated about ten kilometers into the land, killing over twenty thousand people. The water leaped over the seawall, disabling the plant's backup generators and preventing them from cooling down the reactors. The explosions occurring in these reactors were not from meltdowns. They were explosions of hydrogen that had accumulated inside of the reactors after it had separated from oxygen in the extremely hot water vapor. So the containment structures did their job of holding in the nuclear radiation. Unfortunately, some radiation leaked into the Pacific Ocean.

The accident at Chernobyl was by far the most serious. It directly killed about fifty people. This core meltdown was due to a poor engineering design that used graphite instead of water as a moderator (used in most designs) and a foolish test carried out by the plant's operators.

"According to an authoritative twenty-year study by the International Atomic Energy Agency and the World Health Organization, over time this fallout will most likely be responsible for four thousand deaths among the surrounding population."[2] Yet, as bad as this could be, considering the amount of energy that has been produced by nuclear energy, the danger is miniscule compared to the thousands of deaths caused by coal, petroleum and even hydroelectric energy. It has been estimated that for every nuclear generating plant replacing a coal power plant, hundreds of lives would be saved.[3]

How Nuclear Energy Works

In the 1930s scientists discovered that when a neutron is shot into the nucleus of certain unstable atoms such as uranium-235, it splits into two smaller elements, e.g. barium and krypton, called "daughter elements".

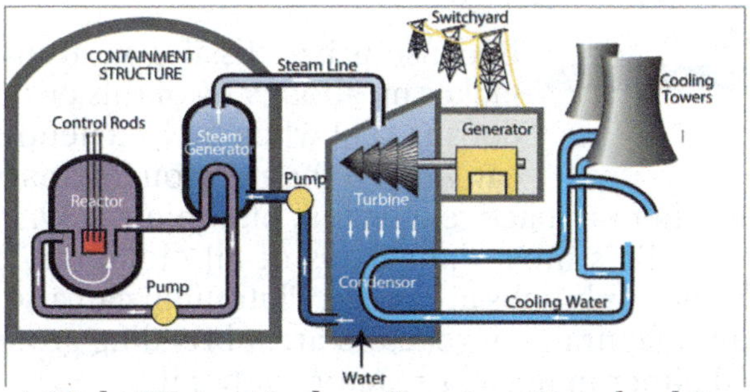

A Light Water Nuclear Reactor: Standard Design

When an atomic nucleus is split, it releases a tremendous amount of energy plus two or three additional neutrons, which are then free to smash into nearby atoms in a chain reaction. The mass of the products of this reaction is slightly less than that of the original material. In this tiny difference in the mass a tremendous amount of energy is released — almost a trillion times as much as the difference in mass. This follows Einstein's equation $E = MC^2$. "E" is the energy produced, "M" is the mass being converted into energy, and "C" is the speed of light, 300,000 kilometers per second. When you square that large number (300,000 X 300,000), you get an unbelievably large number — almost a trillion. Nuclear energy is concentrated energy, producing about two million times as much energy per gram as the energy released from breaking chemical bonds in the combustion of coal or natural gas.

Just as in coal and natural gas plants, the heat energy produced in nuclear reactions is used to heat water to produce steam. Because steam occupies 1,700 times as much space as liquid water, it exerts a tremendous force that is used to spin a turbine and an electromagnet. When the magnet spins around, its north and south poles rotate. This changing magnetic field produces electricity.

Over 99% of the uranium found in uranium ore is U-238, a form of uranium that can't be used directly, because it's not fissile. It can't be split as U-235 can. In order to make fission work, some of this U-238 must be removed from the mix to raise the percentage of U-235 to about 3%. This process, called enrichment, is complicated and costly. Thus, the U-235 isotope is rare and expensive.

In addition to the expense of enrichment is the problem of keeping the water used in traditional light water reactors at the pressure needed to keep it from turning into steam at 100°C. It's like a pressure cooker. If the pressure is accidentally released, it can explode. Water is the moderator, slowing down the neutrons so they can more easily split the U-235 nucleus. If this water accidentally escapes as steam, it can no longer do its job of coolant and moderator. In that case, the reaction would be out of control and could cause a meltdown. This is why light water reactors must be housed in huge, expensive containment buildings.

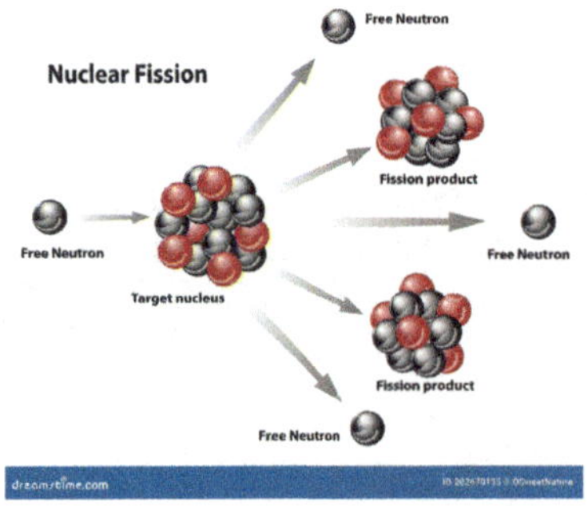

In a traditional nuclear reactor, solid fuel pellets of uranium encased in zirconium are inserted into fuel rods. After about 1% of this nuclear fuel is used up, with each U-235 atom being converted into two smaller "daughter" elements, the pellets swell, cracking their protective coatings. At this point they are unusable and must be replaced, creating nuclear waste.

Because it isn't fissile, an atom of U-238, which makes up about 97% of this enriched fuel mixture, doesn't split when hit by a neutron. But its nucleus can *absorb* the neutron, increasing its weight and changing it into another radioactive element, plutonium-239. Plutonium is fissile and also very toxic. This most dangerous of all elements, plutonium, is the material of which nuclear bombs are made. Plutonium can also be used to power a reactor. So traditional uranium reactors are a breeding ground for plutonium, which can be used either in nuclear reactors or bombs.

Nuclear Energy from Thorium

A totally different nuclear process which was successfully tested for about five years — from 1965 to 1969 — uses thorium, another radioactive element. Because thorium — atomic number 90 — takes longer to decay than uranium, it isn't nearly as radioactive. The half life of uranium is 4.5 billion years. This means that it would take 4.5 billion years for half of a sample of uranium to decay into another element. The half life of thorium is over 14 billion years, about the age of the universe. You can hold a lump of thorium in your hand and it won't hurt you.

When hit by a neutron, the nucleus of a thorium atom doesn't split, but just like uranium-238, it absorbs the neutron. This starts a series of reactions in which thorium changes into another isotope of uranium, U-233, which is fissile. When hit by a neutron, a U-233 nucleus splits into two smaller elements, just as U-235 does. This reaction releases a huge amount of energy plus two free neutrons which, in an enclosed space, can smash into more U-233 atoms in a chain reaction.

Thorium is four times as abundant as uranium. Even better, it doesn't need the expensive enrichment process required by uranium. Therefore, 100% of the thorium that is mined, compared with 1% of uranium, can be used for fuel.

Because thorium is not used as a solid fuel but instead is dissolved in molten salt, 99% of it can be used in the reaction, compared with only 1% of a solid uranium fuel pellet that swells and cracks when its daughter elements are produced. This means that thorium can produce electricity about a thousand times more inexpensively than uranium.

Instead of water, the molten salt in which thorium is dissolved is the coolant in this reaction. Salt has a much higher melting point than water. It becomes a liquid at about 700°- 1,000°C. The salt remains a liquid at these elevated temperatures and, serving as a coolant, absorbs the heat from the reaction. This system doesn't need to be pressurized, as water does in a light water reactor. So no containment structure is needed. Its power plant is thus much smaller and less expensive than a light water uranium reactor. In the event of a power failure, the safety system works automatically. Not requiring human intervention, it is foolproof.

An accident such as the one at the Fukushima Daiichi plant — where the back-up diesel generators failed — could never happen with a thorium molten salt reactor. Beneath the system of pipes through which the molten salt circulates is a "plug" of salt kept solid by a fan blowing cold air on it. If the power should ever fail or is turned off, the fan will stop running, and this plug will melt, allowing the hot molten salt mixture to drain out into larger tanks below. At this point the reaction will cool down and stop, because the atoms are now too far apart to support a chain reaction. This type of reactor would never experience a meltdown such as those which occurred at the Chernobyl, Fukushima and Three Mile Island nuclear plants.

Thorium is so abundant in the Earth's crust that we will never run out of it. A lump of thorium about the size of a tennis ball could provide all of the energy a person needs for a lifetime: for transportation, heating, cooking, light, etc. This would cost a small fraction of a person's current estimated energy cost of $4,657 per year.

The low-cost electricity this reaction would produce could be used to split water molecules to produce hydrogen gas economically. We could thus have hydrogen-fueled automobiles that wouldn't need large batteries. When hydrogen burns, the only product is water vapor. The low cost power obtained from molten salt reactors could be used in desalination plants to provide all the fresh water people need. Desalinated water could transform Earth's deserts into fertile farmland. This is not a pipe dream.

Just one thorium mine could provide all of the world's energy needs for a year. Hundreds of thorium mines have already been discovered. Because thorium is a waste product of the mining of rare earth metals currently used for electronics and for the magnets used to power wind generators, there are currently piles of thorium sitting around near these mines.

By now you might be wondering why, if thorium has so many advantages over uranium, it isn't being used instead to produce electricity. There are a couple of answers to this question. First, the initial research leading to the development of nuclear power occurred during World War II. The leaders of the United States feared that the Nazis might construct a nuclear weapon. At the conclusion of the war, the focus changed to peacetime uses of nuclear power. Uranium was the fuel of choice for leaders who anticipated the possibility of having to use a bomb in some future war. Some scientists believe that if nuclear technology had been developed during a time of peace, thorium, which cannot be used for bombs, would have been the nuclear fuel of choice.

A thorium reactor that ran for almost five years at Oak Ridge Labs in Tennessee was invented by Dr. Alvin Weinberg, the same man who invented the light water reactor that uses U-235. Believe it or not, when this brilliant man expressed his concern over the safety of the light water reactor, he was fired and the thorium project was canceled. But the nuclear accidents at Chernobyl and Fukushima, along with the realization that wind and solar power are not the answer to the world's energy needs, have inspired a renewed interest in thorium as a nuclear fuel.

Thorium Power Plant in the Netherlands

But some technical problems must be worked out before thorium on a commercial scale becomes a reality. One problem is that the molten salts in which thorium is dissolved are corrosive at higher temperatures. An alloy containing mostly nickel to house the reactants has been developed, but they are still taking the bugs out of this. Another glitch is that the element protactinium, the intermediate element into which thorium changes before becoming uranium-233, tends to absorb some of the neutrons needed to keep the reaction going. Therefore, it must removed from the mix at a certain point in the reaction. These problems are being solved. Experimental thorium molten salt projects are underway in China, India, the United States, Denmark and the Netherlands.

Uranium-235 can also be dissolved in molten salt, and the resulting reaction is simpler. The only drawback to this system is the fact that uranium-235, which must be enriched, is much more expensive and rare than thorium. But, instead of only 1% of the fuel being used, this type of reactor can use almost 100% of the fuel, making it much more economical. This type of reactor can also use some of the nuclear waste that couldn't be used in the past.

The first successful use of a molten chloride reactor (MCR) using uranium-235 was recently accomplished at the Idaho National Laboratory in 2025. This is the culmination of experiments that have been conducted since 2020. "The molten chloride fast reactor represents a paradigm shift in the nuclear fuel cycle, and MCRE [molten chloride reactor experiment] will directly inform the commercialization of that reactor," announced Jeff Latkowski, program director of the MCRE . [4]Commercial operations are planned to begin as early as 2030.

"Molten chloride reactors are exciting, because they offer the best blend of fuel efficiency, safety and long-term sustainability," explained research scientist Jacog Yingling. These reactors, not much larger than a tractor trailer container, can be safely used to power ships as well as for smaller communities.

Could nuclear fusion fulfill our energy needs?

Fusion is a nuclear reaction taking place in the sun's interior to produce the energy that we receive on earth. In a fusion reaction atoms of deuterium (H-2) and tritium (H-3) are squeezed together to produce helium. Deuterium is a form of hydrogen that contains a neutron. It is twice as heavy as a typical hydrogen atom. Tritium has two neutrons.

Each proton has a positive charge. Like charges repel each other. So the positively charged protons, repelling each other, must be pushed together under extremely high pressure and high temperatures to fuse together to make a helium atom. When deuterium and tritium are pushed together closely enough, the nuclear strong force, which pulls protons together, can overcome the opposite force, electromagnetism, that is trying to push them apart at the same time. Inside every atom there is a war going on between these two forces.

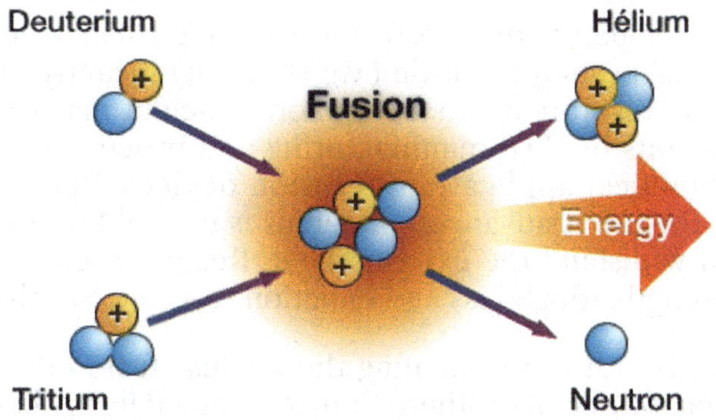

Because of the extreme heat and pressure required to do this, nuclear fusion will be extremely difficult to pull off. It is impossible to find a reactor shell that won't melt from the temperatures of millions of degrees needed in this reaction. The reactants must be held together by powerful magnetic fields. The energy produced when these atoms fuse together must be greater than the energy needed to do this. Otherwise the process wouldn't be practical for commercial use.

Lately there have been some promising developments in fusion technology. Experimenters at the Massachusetts Institute of Technology (MIT) tested a new type of reactor shaped like a sphere. Rather than trying to join hydrogen atoms together, this process would use heat from lasers to fuse hydrogen and boron to produce carbon.

Meanwhile, in Germany another fusion reactor known as the Wendelstein 7-X Stellerator reactor is being readied for operation. A stellerator is a device that uses magnetic confinement to hold particles together. Operational since 2015, this stellerator uses a complex system of coils to hold plasma (atomic nuclei) together. It has set records for plasma confinement and stability but is reportedly a long way from being able to produce sustainable energy. [5]

Operated by the Lawrence Livermore National Laboratory, the National Ignition Facility (NIF) uses a powerful laser system to compress and heat a tiny fuel pellet, triggering a fusion reaction. In 2022 this experiment reached a significant milestone, reaching "ignition status" in which the energy released in the fusion reaction was equal to the energy required to get it to that point. [6]When the energy produced surpasses the energy expended in getting to that point, fusion can be used commercially.

At Texas A & M University, a team of researchers experimenting with helium at high temperatures are also finding the task daunting. When helium heats up to these high temperatures, it begins to bore holes resembling veins in the metal casing.

Another fusion method called Z-Pinch uses a powerful electric current to compress and heat a thin wire, producing a plasma that undergoes fusion. Recent developments in this technology have led to increased energy gains and improved fuel efficiency. [7]

Magneto-Inertial Fusion (MIF) combines magnetic confinement and inertial confinement, using inertial fields to guide and compress the fuel. Recent research in MIF has focused on improving stability and efficiency with promising results. [8]Inertial confinement is heating a small pellet of fuel under pressure using high-powered lasers.

High-Temperature Superconductors (HTS) are using yttrium barium copper oxide (YBCO) and magnesium diboride (MgB2) to significantly improve the performance of fusion reactors by enabling more efficient magnetic confinement. Recent developments in HTS manufacturing and processing have brought us closer to their practical application in fusion devices. [9]The development of new steel alloys and other advanced materials is crucial for constructing fusion reactors that can withstand the extreme heat. Recent research has focused on improving the strength, toughness and radiation resistance of these materials.

Many other organizations are pursuing the nuclear fusion dream. The largest fusion experiment, located in southern France, is called Iter Tokamak. Thirty-five nations are participating in the construction of this massive doughnut-shaped fusion generator.

Iter Tokamak

Iter Tokamak, located in France, is by far the most complicated machine ever built as well as the most complex scientific endeavor in human history.[10] This machine has finally been completed, but they don't expect it to be fired up until 2039 at the earliest. To hold the extremely hot hydrogen and helium atoms together at 150 million degrees, large superconducting magnetic coils are used. Strangely, in order for the magnets to be superconductive, their temperature must be held at a frigid -269°C. So some parts of the machine are exceptionally cold, while others are exceptionally hot. Despite sixty years and billions of dollars spent on this research, producing fusion on a commercial scale seems to be decades away.[11]

Hydroelectric Power

Hydroelectric power uses the downhill force of running water from gravity to spin the blades of a turbine that turns an electromagnet. The changing magnetic field created by the rotating north and south magnetic poles in the magnet is converted to electrical energy. Hydroelectric power is 90% efficient, making it the most efficient energy source available. But the initial construction and start up costs of a hydroelectric power plant can exceed a billion dollars. So it's difficult for third-world nations to get the funds needed to build hydroelectric projects. "In 2021, hydroelectricity accounted for about 6.3% of total US utility-scale electricity generation and 31.5% of utility-scale renewable electricity generation."[12]

Grand Coulee Dam

Yet hydroelectric power plants have their downside. They can cause great environmental damage by flooding large land areas on which natural plants and animals exist as well as converting river valleys once used for agriculture into virtual wastelands.[13] The presence of a new dam can even affect the migration patterns of fish.

Hydroelectric dams can also be dangerous and life threatening. The worst disaster in the production of electrical energy occurred in 1975 when a dam built in China twenty years earlier was unable to accommodate a huge amount of water from a typhoon that had dumped over 40" of rainfall on the area within 24 hours. The sluice gate couldn't take the water fast enough. A huge wave crested over the dam and rushed downstream at 50 km/hour (31 mph) to create an instant flood. Approximately 26,000 people died from this flood, and another 145,000 died later from famine and disease. Almost six million buildings collapsed in this disaster. The risk from nuclear plant meltdowns pales by comparison to this disaster.

Electric Power from Geothermal Energy

Geothermal energy — heat (thermal) from the earth (geo) — is another sustainable alternative to the combustion of carbon-based fuels. Beneath the earth's crust, the average depth of which is anywhere from 15 to 20 km (9.3 – 12.4 miles) lies the mantle, which is extremely hot, over 1,000° C (1,830° F). This is a largely untapped source of heat. The deeper you go in the crust the higher the temperature is.

Heat deep in the earth's crust is generated partly by the decay of radioactive elements in the lower crust and mantle. Some parts of the water table under Earth's surface are heated by hot rocks close to concentrations of melted rock, called magma, that has been pushed up from the earth's mantle. Sometimes this water is even superheated to steam, which is used to spin turbines connected to electrical generators.

Although geothermal energy isn't available everywhere, there are still many locations where one doesn't have to drill more than about 1.5 km (0.9 miles) to reach a sweet spot. In the past this was unheard of, but recently, as oil-drilling technology has improved, the depth to which shafts can be drilled is much greater. The average depth of an oil well has increased from about 1,830 meters (6,000 ft), to as much as 12,200 m, or 12.2 km (40,000 ft). The ability to drill deeper than ever before has significantly increased the number of locations where geothermal energy can be used.

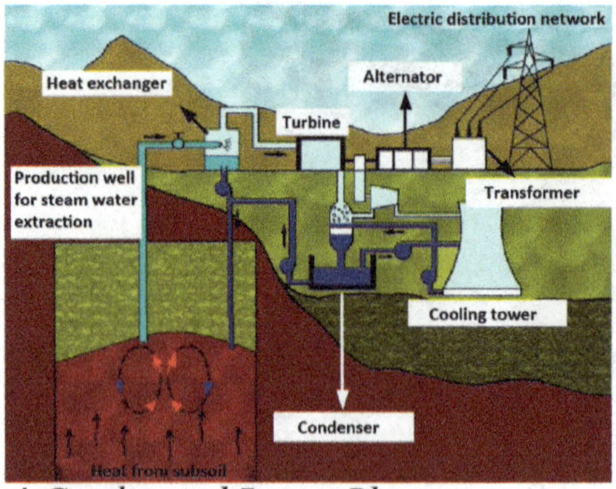

A Geothermal Power Plant Design

One such operation in the US is a plant north of San Francisco called the Geysers. This plant produces about 1.5 GW (billion watts) of electrical power. There are over 20 geothermal installations in Nevada, which, while producing much less power per unit, still add up to about 565 MW (million watts) of electricity.

Over thirty nations are producing geothermal energy. Geothermal heat left over from electrical generation can also be used for directly heating buildings. Iceland, where there are many volcanoes, uses geothermal heat for heating huge greenhouses, in addition to generating electrical power. Many more such projects are being planned – and not just for places where there are natural hot springs. Geothermal energy systems are adaptable for a number of geological environments.

For example, in southern Saskatchewan in Canada, near the US border, a series of wells were drilled 3.7 km (2.3 miles) vertically and connected by drilling horizontally and breaking up the rocks in between to provide more area for water to circulate. Water run through this system picks up heat and is pushed back up to ground level by convection. In effect water is a heat exchanger. The resulting steam drives a turbine that generates 20 MW (million watts) of electricity, enough for 20,000 homes.

Geothermal energy is clean and exceptionally reliable. It is also economically competitive with other energy sources. Artificial intelligence can optimize this next-generation technology by helping to locate the best spots in which to place these generators.

But there are occasionally problems in extracting heat from the Earth in this manner. A Swiss newspaper article dated December 10, 2009, describing a geothermal project in Switzerland, states: "The authorities in canton Basel City say they will cancel a geothermal energy project, which three years ago caused minor tremors that damaged many buildings. A risk analysis study published on Thursday found that the danger of setting off more earthquakes was too great if drilling at the site resumed." Nevertheless, this is not a problem in most areas.[14]

Drilling technology has improved by a factor of 10 in the last ten years. Many people who have worked in oil fields and understand the technology of drilling can apply their skills and knowledge to geothermal operations. So we are probably going to see many more geothermal energy plants built in the near future. "This growth trajectory positions geothermal energy to play an increasingly important role in meeting global electricity demand while supporting decarbonization goals."[15]

Ground source heat pumps aren't the same technology as geothermal heat. A heat pump is an air conditioner that works in both directions. In the summer it transfers heat from living space to the outdoors. In winter, in above-freezing conditions it transfers heat from the outside air to your living space. A ground source heat pump moves heat through pipes buried underground to take advantage of the fact that the temperature underground is always about 55° F (12.8°C), saving at least 50% on heating bills.

Combined Cycle Gas and Coal Power Plants

A Combined Cycle Gas power plant (CCGT) is a very efficient system in which methane, or natural gas, is burned to produce hot gases. These turn a giant turbine which rotates to produce an electric current. After the first go around, the leftover heat is used to heat water above its boiling point of 100°C, converting it to steam. Occupying a volume over 1,700 times that of liquid water, steam produces a tremendous force that turns another turbine to produce more electricity. These plants are very efficient, increasing the amount of electrical energy that can be produced with the same amount of fuel by about 50%. These power plants can also use biogas, diesel or hydrogen as fuel.

Combined Cycle Coal technology is a power generation method using coal gasification to produce synthetic gas (syngas), which is then burned in a gas turbine to generate electricity. This method is more efficient than traditional coal combustion and produces a higher energy output with reduced CO_2 emissions.

Coal is first ground into a powder and then reacted with steam and a limited amount of oxygen at high temperatures to produce a gas composed of hydrogen and carbon monoxide: syngas. This gas is burned in a natural gas turbine that spins an electromagnet to produce electricity. The leftover heat from this reaction is then used to heat water to steam, which rotates another turbine, producing more electricity. Filters and scrubbers prevent harmful gases and particulates from polluting the atmosphere.

Because it will take several years to ramp up our nuclear energy capability, we will be relying on these advanced coal and gas power plants that produce more energy for the same amount of fuel.

Conclusion

According to the IEA and EIA, the world's demand for energy is projected to increase by 30% to 75% by 2050. This will be due to population growth and rising living standards in developing countries, as well as the growth of Artificial Intelligence. Energy powers just about everything in our modern society: heating, air conditioning, transportation, manufacturing, agriculture, education and communication. For this we need energy that is low cost, dependable and sustainable. Energy lifts people out of poverty, and, as advanced energy technologies are developed and commercialized, the world will enjoy a level of prosperity that was only dreamed of previously.

Now that they've been extensively tested, we know that wind and solar power systems are not renewable and have proven to be undependable. They are not the answer to our energy needs. For the time being we will have to continue to rely on the combustion of hydrocarbons, which can be accomplished more efficiently with combined cycle gas and coal electricity plants.

Exciting new nuclear power technology includes molten salt uranium and thorium reactor designs, which are well on their way to becoming a reality. Nuclear energy has a proven safety record as one of the safest forms of power generation.

In addition, we will see more electricity powered by geothermal systems.

These technological advances will create a tremendous increase in prosperity, providing many more jobs, as we advance into the second quarter of the 21st century.

---

1. Zubrin, op cit., p. 133.
2. Ibid.
3. Op. cit., p. 142
4. Donna Kemp Spangler, "Idaho Lab Produces First-Ever Fuel for Fast Molten Salt Reactor Experiment", *Idaho Business Review,* Dec. 26, 2025.
5. Max Planck Institute for Plasma Physics, 2021.
6. Lawrence Livermore National Laboratory, 2022.

7. SciTech Connect, 2021.

8. Fusion Science and Technology, 2021.

9. Superconductor News, 2021.

10. https://www.theguardian.com/environment/2017/dec/06/iter-nuclear-fusion-project-reaches-key-halfway-milestone

11. Ben Turner, "World's largest nuclear fusion reactor is finally completed. But it won't run for another fifteen years." *LiveScience.com*, July 3, 2024.

12. U.S. Department of Energy, "Hydropower Program"

13. https://www.eia.gov/energyexplained/hydropower/

14. "Future of Geothermal Energy Growth Technology Breakthroughs," *Energy News*, April 8, 2025.

15. Ibid.

Chapter 12

The Final Analysis

> "If you tell a lie big enough and keep repeating it, people will believe it. The lie can be maintained only for such time as the state can shield the people from the political, economic and/or military consequences of the lie. It thus becomes vitally important for the State to use all of its powers to repress dissent, for the truth is the mortal enemy of the lie, and thus by extension, the truth is the greatest enemy of the State." — Joseph Goebbels, Third Reich Minister of Propaganda

Carefully studying and verifying all of these facts, one can only conclude that the climate change hypothesis doesn't stand up to scientific scrutiny. It is a political issue. To capture the moral high ground, a group of the world's elite with the goal of establishing a totalitarian world government usurped environmentalism, a noble cause. At the same time the way they live their lives demonstrates that they themselves don't care about the environment. It has been reported that Al Gore's house uses 34 times as much electricity as the average home. John Kerry, Biden's climate czar, defended his flying all over the world in a private jet by calling it, "The only choice for somebody like me who is traveling the world to win this battle."[1]Private jets emit up to forty times as much carbon dioxide per mile as airline travel.

These people have clearly stated that they want to reduce the world's population from the current 8.2 billion to as low as 500 million (a 16-fold reduction). In 2010 Bill Gates made it very clear in a Ted talk that he believes the world's population should be much smaller. (This quote was mostly wiped from the internet, but it was archived.)

> "One of these numbers is going to have to get pretty close to zero...Let's take a look. First we've got population. The world today has 6.8 billion people. That's headed up to about nine billion. Now if we do a really great job on new vaccines, healthcare, reproductive health services, we could lower that by perhaps ten or fifteen percent."

"A total population of 250-300 million people, a 95% decline from present levels, would be ideal." —Ted Turner, founder of CNN and major contributor to the UN

"The resultant ideal sustainable population is hence more than 500 million but less than one billion." —Club of Rome, *Goals for Mankind*

The globalist elites founded the Club of Rome in 1968. *The First Global Revolution*, a report they wrote in 1991, gave their strategy away in this statement, **"In searching for a new enemy to unite us, we came up with the idea that pollution, the threat of global warming, water shortages, famine and the like would fit the bill."**

The drastic measures outlined in the Green New Deal, [2] that the Biden regime, with the help of the Inflation Reduction Act passed by the US Congress, called for completely replacing "fossil fuels" with "renewable" sources: wind, solar and biofuels by 2050. Their goal was to replace all gasoline and diesel engines with battery-powered cars, trucks, farm tractors and even construction vehicles. Electric utilities would rely on "renewable" generation.

After trillions of dollars have been spent, wind and solar together provide only about 2.8% of the world's electricity. Meanwhile there is more demand than ever for hydrocarbons, which provide 86.7% of our electricity, according to the new, correct method of calculation by the Statistical Review of World Energy (SRWE).

The severe winter storm that hit Texas and nearby areas in February, 2021, demonstrated that not only is the climate not warming but that nearly half of the wind turbines, which had cost Texas residents upwards of $30 billion ($1,000 per resident), froze up and ceased operating. As a result, millions of people had no electricity for several days. Over 200 people died from the cold.

Texas wind turbines

Of course the dark legacy media blamed this situation on problems with natural gas, carefully leaving out the fact that the wind turbines, which had stopped operating, had put more pressure on the natural gas plants than they could handle. And of course, the cold weather was blamed on climate change.[3] "Scientists who study climate change claim that Texas' extreme weather is due to the heating of the Arctic."[4]

A major electricity blackout in Spain, Portugal and some parts of France in April of 2025 caused tens of millions of people to lose power for several hours. This blackout started with a couple of quick power interruptions in southern Spain caused by large solar farms suddenly disconnecting from the grid. This led to a sudden spike in voltage, which the grid couldn't handle quickly enough. Protective systems kicked in, automatically shutting off more power sources to avoid damage. More shutdowns meant even higher voltage and a resultant drop in amperage, leading to the whole system shutting down in seconds.

The sun was especially bright that day, and solar power was cranking out at a high level. This meant that there were fewer traditional power sources operating at that time. The power grid is like a giant seesaw that needs to stay balanced. Power oscillations can also occur in traditional power setups. But they are offset by the greater inertia of heavy, spinning turbines, which provide stability to the system. With bright sunshine the solar component, using inverters instead of turbines, was providing most of the power. This caused instability.

This led to chaos and pandemonium. Subways stopped working. Cash machines were inoperative. Phones and internet signals were cut off. Food in supermarkets was spoiling without refrigeration. Hospitals were thrown off guard, and home medical units were unable to deliver oxygen. Most people fail to consider just how dependent the modern world is upon dependable electricity, but the people in Spain quickly found out. Fortunately this didn't happen in the winter.

At this moment (January, 2026) the continental United States is experiencing the worst onslaught of winter weather in many years. Half a million people are currently without power as the specter of another round of frigid temperatures and snow looms. One of the states hit hardest is Massachusetts, where wind provides only about 1% of its electricity. Massachusetts imports electricity from a hydroelectric generator in Quebec. In view of the circumstances Quebec decided to temporarily keep this energy, forcing Massachusetts to burn oil for heat. This move could be permitted only by suspending EPA regulations, allowing oil to be burned for heat. Massachusetts has been burning 150 metric tons of oil per minute to keep people from freezing.[5]

In Summary

The United Nations Intergovernmental Panel on Climate Change (IPCC) has announced that humans are responsible for the small amount of warming that has occurred since the end of the Little Ice Age around 1800, totally ignoring natural causes of climate change. Without real world evidence to back up this hypothesis, they continue to claim that if their dictates are not followed, the earth's average temperature will increase to dangerous levels in the near future. This, they say — also without any evidence whatsoever — will cause the extinction of many species of plants and animals, extreme weather events and danger to food supplies, human health and economic growth. In the decade from 2013 to 2023 between 5 and 10 trillion dollars (US) was spent to avoid using hydrocarbons. The IPCC estimate of the cost to remedy this "problem" is $2.4 trillion US per year.[6]

Yet many prominent scientists — including Nobel laureates — have expressed serious doubts about these claims. Each year that tipping points come and go the number of skeptics increases. Claiming that the issue is settled, the doomsayers refuse to debate it with other scientists. Instead they call those who question them "climate deniers" or even "science deniers". They've done everything they can to silence voices of disagreement.

One spokesman for human-caused climate change is TV personality and clown, Bill Nye, the Science Guy, who isn't a scientist. Nye, Al Gore and other politicians have even called for jailing climate skeptics under RICO, the Racketeer Influenced Corrupt Organizations Act.

When Tucker Carlson asked Nye in 2017, "To what extent has the rate of climate change been increased by human activities and how much of it was caused by natural causes?" Nye was unable to provide an answer. If the science is indeed settled, the "experts" should certainly know the answer to this important question. NOAA administrator John Knauss was forced to admit, "In spite of all the well-publicized concern about global warming, you must understand that there is still considerable uncertainty among scientific experts about a number of critical factors which determine global warming." [7]

Richard Lindzen, PhD, is one of the world's most respected climate scientists. As the Alfred P. Sloan Professor Emeritus of Meteorology at MIT, Dr. Lindzen has made valuable contributions to climate science. He stated:

> "What historians will wonder about in future centuries is how deeply flawed logic, obscured by shrewd and unrelenting propaganda, actually enabled a coalition of powerful special interests to convince nearly everyone in the world that CO_2 from human industry was a dangerous, planet-destroying toxin. It will be remembered as the greatest mass delusion in the history of the world — that CO_2 , the life of plants, was considered for a time to be a deadly poison."

Since the debut of AI there have been signs that the push to achieve Net Zero is fading. The Trump administration is putting the brakes on its implementation. Yet, if another climate alarmist government takes over — which is likely — and re-establishes these mandates, your life will never be the same again. Your taxes will skyrocket. You will be banned from state and national parks.

You will be told what kinds of food to eat. You won't be permitted to eat meat, because the plan's proponents claim that meat production produces methane. The raising of cattle and other farm animals for meat will be prohibited. The excuse for such action is that livestock emit methane (CH_4 , or natural gas). They never tell you that methane exists in a very tiny amount: just two parts per million (0.0002 percent of the atmosphere. The amount of methane emitted by these farms would have to be multiplied by many times, to ever be such a threat. By subjecting farmers to a multitude of impossible regulations, the world's food supply has been placed in great jeopardy.

The proponents of this plan want all vehicles to be electric, even though hydrocarbons still generate most of the electricity used to fuel them. Does this make any sense? They want you to travel only on railroad lines, by bicycle or on foot. They want to phase out air travel. Every aspect of your life will be micromanaged by unelected bureaucrats. This is true socialism at best, but most likely, totalitarian communism, even though those terms are omitted from their documents.

Consider the fact that none of the leaders of the United Nations or the European Union have been elected. They obviously feel no accountability to the people. Who appointed these people to their high positions in which they dictate to the world what is to be done?

Please note that these measures apply to us peons, not to the elites who are promoting them. They have even called for us to get rid of our pets. One "green" suggested that people shower together.

Even some physicians are on the band wagon. "For a long time there was a notion that the greenhouse effect caused in healthcare settings was an inevitable and unavoidable cost of providing patient care, but we have learned that **reducing anesthetic gas flow** is one of the many ways healthcare can lessen its contribution to the healthcare crisis," declared one doctor. So they would cut down on the amount of anesthetics (such as N_2O) used in operations. Physicians are even being encouraged to add "climate change" as a cause of death on death certificates, just as they did to inflate deaths from COVID.

But, as said before, this is really a political issue. Alexandria Cortez's (AOC's) former chief of staff, Sycat Chakrabarti, revealed in a recent meeting, "The interesting thing about the Green New Deal is it wasn't originally a climate thing at all. Do you guys think of it as a climate thing? Because we really think of it as a how-you-change-the entire-economy thing."[8]

The Secretary General of the United Nations, Antonio Guterres, a self-proclaimed socialist, affirmed, "We need to make sure we reverse the trends. It's now clear we are coming to a point of no return. To spend those trillions of dollars and not use this occasion to reverse the trends and massively invest in the Green economy will be an unforgivable lost opportunity."[9]The Rockefeller Foundation announced its partnership with the World Health Organization (WHO) by contributing five million dollars "to cultivate global networks for pathogen detection and strengthen pandemic preparedness capabilities, including broadening surveillance for diseases worsened by rising temperatures and extreme weather." [10]One key word here is "surveillance". You can draw your own conclusions.

To help pay for climate mitigation, the world's central banks have printed and put into circulation trillions of dollars. It is a basic tenet of economics that putting more money into circulation creates inflation, devastating peoples' savings and retirement accounts. Inflation is the hidden tax. We need only to look at what recently happened in Venezuela. The rate of inflation that just occurred there under a socialist dictatorship was as high as 1.6 million percent. Before allowing our lives to be taken over in an experiment to "prevent the world from ending", we need to ask, "Is the earth really headed for disaster if we fail to go through with these drastic measures that will shake our way of life to its very foundation?"

As a voter or policy maker, you need to know as much as possible about this issue. This is the only way you'll know which candidates for office to support: those who believe we can change the climate by reducing CO_2 or realists who recognize the fact that climate change is not a crisis.

As stated before, the entire "climate change" scenario that was cast upon the world in the 1970s — first as a global cooling crisis and then as a global warming crisis — is playing with the world's need for energy, without which our civilization will revert to the Dark Ages. In the not-too-distant past energy was provided mostly by animals and by human slaves as well as by burning wood. Wood, as well as coal, was used in the first steam engines at the beginning of the Industrial Revolution in the early 1800s. The first oil well was dug in 1859. Since then our use of energy has changed dramatically.

As the 21st century progresses, the world's people will need more and more energy, on which all life depends. The world currently relies heavily on carbon-based fuels, which still provide over 86% of the world's electricity. After all the trillions that have been spent, wind and solar together provide under 3%.

There are many health hazards associated with coal mining. But every form of energy production has drawbacks. Alex Epstein noted, "When evaluating any issue, we must look at the full context, carefully weighing the benefits and side effects of every option." Our shortsighted politicians, who obviously have never studied the issue, have utterly failed to do this.

This oil well, from which 100% organic material is pumped out of the ground, requires about 45 -90 square meters (500 to 1,000 ft^2. The petroleum is then transported safely through pipelines to a refinery to be processed into usable compounds that are made into over 6,000 products for society. Hydrocarbons are the basis of modern transportation, including aviation, commercial ships, and military vehicles.[11]Steel cannot be made without coal.

It would be impractical and inhumane to hold back from the world this inexpensive and dependable source of energy that people desperately need **now** for heating, cooling, transportation, agriculture and food storage. For those in power to deny the world's people access to carbon-based fuels without offering a viable and sustainable alternative — such as nuclear energy — would wreck the entire world economy. But this is what the founder of the UN IPCC, Maurice Strong, said he wanted to do when he asked, **"Isn't the only hope for the planet that the industrialized civilizations collapse? Isn't it our responsibility to bring that about?"** [12]Withholding these energy sources from people would also sentence millions to death from starvation, freezing and disease.

Most everyone agrees that in order to reduce pollution cleaner methods of producing energy other than burning hydrocarbons need to be found. Most climate alarmists will have nothing to do with nuclear energy, which has been made almost impossible by excessive regulations. Thus, the fact remains that **at the present time** the world's people rely almost exclusively on these dependable, low-cost sources of energy that have enabled humanity to thrive. No one can deny that affordable, abundant energy from coal, natural gas and petroleum has made life for billions of people much better than it was when all of our energy came from burning wood, human slaves and beasts of burden for farming, heating and transportation.

In the 1970s we were told that in another twenty years the world would run out of petroleum. Since that time more and more oil deposits have been discovered. There are now about 65,000 known oil fields, many of which are gigantic. Over 900 of these contain at least five hundred million barrels (about 79,000,000 cubic meters) each. The oil companies don't want the public to know just how much petroleum there is and how cheaply it can be pumped out of the ground. Nevertheless, most people would agree that we shouldn't continue to burn gasoline and diesel when cleaner, cost-effective and sustainable alternatives, such as nuclear and geothermal energy are readily available.

As Christopher Monckton so aptly stated, "Efforts to make a nonexistent problem go away would hurt the poor the hardest."[13] It is immoral to withhold from third-world nations the same opportunities that boosted the richer nations to their current level of affluence. Energy now provided by coal, natural gas and petroleum can be gradually replaced with safe nuclear reactors.

If human-caused CO_2 emissions — responsible for less than 5% of the slight increase in the trace gas carbon dioxide — are not significantly warming the earth, then it follows that the trillions of dollars already spent and planned to reduce these emissions will make little or no difference in the resulting temperature. This money will have been totally wasted.

This includes huge subsidies for the wind, solar and biofuel industries, administration, propaganda campaigns, and tax breaks [welfare] for the affluent who install solar panels on their homes and buy electric vehicles. These wasted trillions of dollars could have been spent on solving the earth's real environmental problems, such as recycling plastics, metals and paper, curbing deforestation, and removing pesticides, dioxin and nitrates from our water supplies, to name just a few. Just the $3.4 billion spent each year to pump carbon dioxide into the ground (sequestration) alone could be used to provide clean water each year to a hundred million people who don't now have it, saving millions of lives.[14]

Several of the world's leading economists contend that the funds now being wasted could be much better invested in research and development of cost-effective, sustainable energy alternatives. When these resulting innovations come about, as they ultimately will, they'll quickly be put in place by the free enterprise system as has happened throughout history. Kerosene replaced whale oil in the 1860s. Automobiles replaced horses in the early 1900s, and the catalytic converter solved the air pollution problem posed by automobiles. The Green Revolution in agriculture saved millions from starvation in the 1970s.[15] These improvements were made under capitalism and free-enterprise, not through mandates. Such improvements have **never** been made under communism.

Coal will continue to dominate the world's sources of energy until an adequate amount of safe nuclear energy — either fission of fusion — becomes commercially available. With our present technology coal is absolutely essential for the production of steel and cement. Without steel vehicles and other machines cannot be produced.

In contrast, wind turbines work only when the wind is blowing just right. They must be backed up 100% of the time by coal, natural gas or nuclear power plants. The same is true of solar voltaic panels, which don't work at night, in the winter or in cloudy weather. It has become increasingly obvious that the three "renewables" — wind, solar and biofuels — being pushed on us by the climate alarmist cabal — have a limited application and can never replace carbon based fuels. Trying to force these on people with subsidies, cap-and-trade schemes, regressive taxes and draconian mandates will destroy the world's economies and end up killing millions of people.

Further evidence that these "renewables" are ineffective substitutes for carbon based fuels is the fact that for several decades these alternatives have been subsidized by government grants, some of which go back to the 1970s. Without such subsidies these industries would have failed long ago.

Since the institution of "green" taxes the cost of electricity in Great Britain has risen by an appalling 268%. A million elderly British have been struggling to keep warm, and each year over 2,700 of these people have died from the stress of extended periods of cold weather. "There are hundreds of thousands of people who are struggling at the margin of survival...Over 15% of the population have an income below the poverty line...If [an increased] carbon tax comes in, these are the people that are most vulnerable," noted Dr. Sean Healy, head of Social Justice Ireland.[16]

Neil Chatterjee, the former chairman of the Federal Energy Regulatory Commission, noted, "We have prioritized ["Green Energy"] over reliability, and sadly it's going to take catastrophic events to recalibrate things."

In the last few years activists have put more and more pressure on politicians and the public with carefully planned demonstrations and climate lawfare. Activists working with Extinction Rebellion glue themselves to statues and throw tomato soup at priceless paintings in art galleries. Across Europe students occupy schools and universities, shutting them down in month-long campaigns. Another activist group called Tyre Extinguishers claims to have disabled over 11,000 SUVs by removing the air from their tires. In their recruitment of climate warriors, they claim to have no leaders, an example of 5th-generation warfare tactics.

In January of 2026 tens of thousands of residents of Berlin, Germany, lost electric power for several days due to arson. A climate activist group called Vulkangruppe (the Volcano Group) took responsibility. Cell phones and a commuter train were also affected. This criminal action is going too far and shows the lack of compassion these people have for others. At least one person died as a result of this action.

Extinction Rebellion Demonstration; Breitbart.com

An organization called the Climate Emergency Fund boasts that their mission is to "raise funds for and make grants to the disruptive, nonviolent climate movement." So far they have given away at least $5.1 million to 44 climate activist groups. The *New York Times* reported that even oil company heirs are contributing to this organization: for example, Aileen Getty, heiress of the Getty fortune. The groups receiving this money engage in vandalism, which is illegal. According to Caitlin Sutherland, executive director of the conservative watchdog group Americans for Public Trust, the Climate Emergency Fund "has quickly become the ATM that radical environmental activists turn to in order to fund their latest disruptions,"

Their website states: "We supported the NASA scientist who went viral sounding the alarm on the climate emergency by chaining himself to a Chase Bank and the bold activists protesting at major sporting events and museums." Members of one of their recipients, "Just Stop Oil", threw tomato soup onto a Van Gogh painting worth $84 million. They also blocked traffic in the United Kingdom. Another group, "Declare Emergency", organized a paint-smearing campaign across a sculpture display case at the National Gallery of Art in Washington, DC. None of these measures would be necessary if a climate emergency really existed.

The alarmists tell us that it is our duty to make these sacrifices for future generations: our children and grandchildren. But if we bankrupt ourselves and wreck our economy, what sort of legacy are we leaving them? It would make far more economic sense to simply adjust to the slight warming that has occurred since the Little Ice Age, a period of time that some scientists say was the coldest in the last 10,000 years.

> "Instead of implementing economically destructive policies of more taxes, regulations, and subsidies, federal and state policymakers should remove government-imposed barriers to energy innovation. Allowing all forms of energy to compete equally in a free market will enable the U.S. to make tremendous strides both for the economy and for the environment."[17]

When push comes to shove, since sane people would never volunteer to give up their precious liberty and the wealth achieved through the free enterprise system, only a fascist or communist totalitarian government backed up with big guns will ever be able to achieve such a feat. Don't think this couldn't happen. It has been said that those who haven't learned from the past are doomed to repeat it. We only have to remember how the Nazi government in Germany controlled its people with similar lies that led to war and the deaths of millions of people. The same tactics were used in power grabs by twentieth-century communist governments, which were ultimately responsible for the deaths of over a hundred million people. We need to learn from history and also teach these facts to our children before it is too late.

In order to persuade people to do things that are repugnant to them — like accepting a worse standard of living and giving up their freedoms — a lot of fear and guilt must be heaped on them. Most people recoil at the thought of war. So throughout history corrupt leaders have softened up the people's resolve to avoid war by using corrupt, state-run media to manufacture a barrage of false reports about the "enemy".

On Nov. 5, 2019, an article in Bloomberg reported that 11,000 scientists — so-called "experts" — say we're having a climate emergency and the world needs fewer people. In this case the enemy is "climate change," and we are told that we're in a war.[18]It is indeed true that we are in a war: an information war. The best thing we can do is to learn as much as possible about this issue and be ready to ask questions of policy makers — your representatives and other government officials — and people you know who need to learn this information.

If you still have any doubts about whether We the People should follow the dictates of globalist elites, such as the World Economic Forum (WEF), consider the fate of several nations that tried to do so. The companies that belong to the WEF are committed to following its energy policy, abbreviated ESG. (Environmental,

Social and Governance.) The unelected leaders of the World Economic Forum, currently chaired by Larry Fink, have set themselves up as world leaders. They've gotten away with this, because the world's largest corporations have banded together under the WEF banner. ESG requirements are a set of standards for a company's operations that "socially conscious" investors (banks) use to screen potential investments. It is now clear that ESG investing produces inferior returns quite aside from the fact that these investment criteria make no sense to begin with.

Nations compete for high ESG scores to obtain loans from these banks.[19]The first country to be destroyed by this insane policy is Sri Lanka, a small island nation located off the coast of India. In 2019 Sri Lanka was prospering. But in order to obtain a high ESG score, their leadership was sold on ideas published in a WEF article by Joseph Stiglets that if they would stop using "fossil fuels" and artificial fertilizer, they'd be rich by 2025. Instead of this, the opposite happened. They ran out of both fuel and fertilizer, and because their crops were ruined, these starving people revolted against their government. Even the *Wall Street Journal* called the country's failed "Green New Deal" a "human disaster" that decimated its economy and monetary system. Rajapaksa, the prime minister, was forced to resign by email as he was fleeing from the angry mob. This opened up a power vacuum that allowed the WEF to quickly put one of their own people into a position of leadership there. Sri Lanka was the "canary in a coal mine".

A similar plight devastated the country of Ghana, located on the west coast of Africa. Several years ago Ghana was doing well with one of the fastest-growing economies of the world. In 2015 this headline was published: "World Bank Approves Largest Ever Guarantees for Ghana's Energy Transformation." The World Bank promised Ghana, "technical assistance for energy sector reforms and the drafting of a new renewable energy law." As a result of this "guidance" by people in NGO's posing as experts who probably had little, if any farming experience, Ghana suffered a complete shutdown of its electricity and (in some places) water. They also experienced serious food shortages.

The government of the Netherlands (with an ESG score of 98.7) is forcing its farmers to get rid of at least 30% of their livestock (such as sheep and dairy cows). The farmers say that this action will put some of them out of business. The government has taken steps to confiscate about half of their land, which has been held in these farm families for generations. This has led to protests and riots. Farmers in Italy, Spain, Germany and Poland are rising up in support of these farmers, knowing that they will probably be next. When interviewed by Epoch TV in 2022, Thierry Baudet, a member of the Dutch parliament, stated, "We have eight years [Agenda 2030] to fight the Globalist takeover...This is the single most existential fight in the history of civilization."[20]

The country of South Africa, with an ESG score of 91, also experienced severe food shortages. They switched to "renewables" in 2015. So the same story has been repeated over and over and could also take place in the United States, Canada and Australia. If, after the end of the Trump administration in 2029 a new administration re-establishes climate mandates such as electric vehicles, curtailing of gas appliances, etc., and if people haven't wakened up to what has been happening in other parts of the world, the United States will follow on this same path. We all must refuse to comply with the ridiculous mandates imposed on the world by policymakers who have no idea what they're doing. Sadly, because the globalists control the media, most people hear just one side of these issues.

The World Economic Forum and their operatives want to withhold the fertilizer that's absolutely necessary for farmers to produce food for the world's people. Without nitrogen fertilizer derived from natural gas agriculture would support only about 1-5 billion people. Withholding fertilizer would lead to the genocide of **billions** of people. Massive human sacrifices are required to meet the CO_2 stabilization goals of the Kyoto Protocol, which requires an immediate 60% - 80% reduction in carbon dioxide.

When you realize that IPCC climate change "science" is not based on observation and measurement needed for a true scientific investigation, but on pseudoscientific computer models and baseless assumptions, you will want to share these facts with others. We now find ourselves in what is being called "Fifth-Generation" warfare, in which psychological operations (psyops), designed to influence our perceptions are being implemented. It's time to call them out. You can choose to take action to fight back or roll over and allow yourself and others to be victimized. In another generation or so people will probably look back with wonderment on this strange period of history.

NOTE TO THE READER:

It's great that you've read this book to the end. Hopefully the facts it has presented have provided tremendous insight and value to you.

If this is the case, it would be greatly appreciated if you left a short review to let others know how they can also benefit by reading it.

Feedback also helps us learn what is most important to our readers. We welcome your comments and would love to hear from you.

Thanks!

Lynne

Faraday Science Institute (FaradayScience.org)

Email address: contact@faradayscience.org

---

1. Sam Dorman, "Kerry family private jet has taken 16 flights this year alone, records show", Fox News, August 5, 2021
2. https://www.congress.gov/bill/116th-congress/house-resolution/109/text
3. https://www.matchcollegiate.org/2021/04/25/the-texas-ice-storm-of-2021/
4. Ibid.
5. Mike Urban, "You'll Never Trust Green Energy Again After Hearing This", Mike Urban YouTube channel, Jan. 27, 2026.
6. F. William Engdahl, "The Dark Story Behind Global Warming aka Climate Change", Oct. 16, 2018, journalneo.org
7. Tony Heller, RealClimateScience, 5/14/2023

8. Jack Crowe, "AOC's Chief of Staff Admits The Green New Deal is Not About Climate Change", Yahoo News, July 12, 2019.

9. Oliver Milman, "Antonio Guterres on the Climate Crisis: 'We Are Coming to a Point of No Return,' " The Guardian, June 11, 2021.

10. Press Release, "The Rockefeller Foundation and World Health Organization Announce Partnership to Expand Global Pandemic Preparedness in Era of Climate Change," TheRockefellerFoundation.org, May 23, 2023

11. Ronald Stein, "Is It Unethical To Buy A Vehicle Powered By A Lithium Battery?", Issuesinsights.com, June 13, 2022

12. Matthew Ehret, *The Canadian Patriot.*

13. Christopher Monckton, "Kicking the Poor in the Teeth: the Cruel Cost of Needless Mitigation," speech at the Heartland Institute, July 2, 2015

14. Joanne Nova, The Skeptic's Handbook, 2009

15. Ibid.

16. "Carbon Tax To Hit Lowest Earners Hardest, Says Social Justice Ireland,"https://www.thegwpf. om/carbon-tax-to-hit-lowest-earners-hardest-says-social-justice-ireland/, Jan. 4, 2019

17. https://www.heritage.org/energy-economics/report/assessing-the-costs-and-benefits-the-green-new-deals-energy-policies

18. Eric Roston, "Earth Needs Fewer People to Beat the Climate Crisis, Scientists Say," Bloomberg, Nov. 5, 2019

19. https://www.future.enverus.com

20. Roman Balmakov, "Facts Matter," Epoch TV, July 13, 2022

Appendix 1: The Climate Change Religion*

* Based on the ideas of Michael Crichton, author of *Jurassic Park* and *The Lost World*

Eden: The world before the Industrial Age

Original Sin: Burning "fossil fuel" for energy

Fall from Grace: Eating from the Tree of Knowledge (looking for the truth)

Sins Against the Climate: Heating, Air Conditioning, Petroleum-fueled Vehicles (Cars, Trucks, Planes, Trains), Eating Meat, Having Pets, Having Children

High Priest: Al Gore

Earth Day, April 24: Religious Holiday

Heresy: Climate Change Skepticism

Sacrifices to the Climate God: People in the Third World who are Prevented from Developing Low-Cost Energy from Hydrocarbons

End Times: The Climate Catastrophe

Tithes: Higher Prices, Higher Taxes and Subsidies for "Renewables"

Salvation: Net Zero at a Tremendous Cost

Appendix 2: Some Talking Points

A Summary of the Climate Change Situation

1. Climate change is a political issue, and the IPCC is a political organization.
2. The percentage of carbon dioxide in the atmosphere has risen mainly because the warming ocean has been releasing it.
3. The earth has been recovering from the Little Ice Age, believed to have been the coldest period in the last 10,000 years.
4. The residence time of carbon dioxide in the atmosphere is 5-10 years, not hundreds or thousands of years as claimed.
5. There is no proof or even evidence that carbon dioxide is Earth's temperature control knob.
6. The "Urban Heat Island" effect has falsely inflated recent temperature readings.
7. We have no idea how much the earth's mean (average) temperature has risen in the last 200 years.
8. Earth is not a greenhouse. The main "greenhouse gas" is water vapor. Carbon dioxide is a minor "greenhouse gas" making up about 0.04 of one percent of the atmosphere, 420 ppm. Consisting of only about 2 ppm, methane is not a threat to the environment.
9. So far taxpayers and consumers have been bilked out of trillions of dollars (US) to feed the trillion-dollar wind and solar industries, offsets, climate taxes and subsidies.
10. Agencies such as the National Oceanic and Atmospheric Administration (NOAA) and the National Aeronautics and Space Administration (NASA) have lowered past temperature records and raised recent temperature records, creating the appearance of rising temperatures. **This would not be necessary if there was a genuine climate crisis**.
11. Thousands of people have taken advantage of these lies to get filthy rich from climate change mitigation schemes.
12. Proxy temperature data from the last million years of geologic history clearly show that changes in temperature occur **before** changes in carbon

dioxide levels.

13. "Globally, weather-related losses have not increased since 1990 as a proportion of GDP. They have actually decreased by about 25%.
14. Scientists who have dared to disagree with those in power about climate change have lost their research grants and sometimes even their jobs. They've been ostracized, vilified, censored and even threatened with prison.
15. In order to build enough solar panels and wind turbines to meet the Net Zero goals of the United States, 375,000 square miles of land area will be required. These are devastating to wildlife and to the environment in general.
16. Offshore wind turbines have killed hundreds of whales and dolphins and are wrecking the fishing industry. So-called "environmentalists" have turned a blind eye to this unforeseen tragedy.

Questions for People who are Worried About Climate Change

1. How is it that, in the 1960s and 70s, when carbon dioxide emissions were rising rapidly, the earth was actually cooling?
2. During the Medieval Warm Period, 1,000 years ago, the earth's average temperature was about two degrees higher than it is now. Humans weren't burning fossil fuels at that time. What caused that temperature increase?
3. Experience has shown that wind and solar-generated energy are undependable, devastating to the environment and three times as expensive as energy from hydrocarbons. What source of energy do you think we should use to replace hydrocarbon fuels?
4. How can solar and wind energy power all of the cars, trucks, farm tractors and heavy construction vehicles currently used to get things done?
5. If the global warming hypothesis is valid, why have agencies such as NOAA and NASA found it necessary to change temperature records by lowering earlier temperatures and raising more recent temperatures? (Yes, they have admitted to doing this.)
6. What is the greatest amount of time that current batteries can store the energy needed to prevent blackouts when the wind isn't blowing and the sun isn't shining?
7. Did you ever hear about the "Climategate" scandal (in 2009 and 2011)?
8. How is it that Al Gore, one of the prime spokesmen for climate change, increased his net worth from $2 million in 2001 to well over $300 million?
9. Since most electric vehicles are charged with electricity produced by "fossil fuels", how does this help avert climate change?
10. Why are Germany, Spain and other European countries which were once dedicated to going "green" starting up their coal-fired electric generators

again?

11. When all the vehicles on the highway are electric, where will the electricity to power them come from?

12. From 2005 to 2017, a 12-year period, no major hurricane hit the US coast. This was the longest pause in hurricane activity in recent history. How can anyone say that there are more hurricanes now then there were when the percentage of carbon dioxide was lower?

13. If wind and solar energy are free, why has the cost of electricity in the United Kingdom and Germany, which have invested heavily in wind and solar, almost tripled?

14. What is the main "greenhouse gas"?

15. Why does Big Oil love wind and solar power?

16. In efforts to reach "Net Zero", if we stop producing synthetic nitrogen fertilizer, which relies on natural gas, what fertilizer will be used in its place to fulfill the UN sustainable development goal of no hunger?

17. There are over 6,000 products made from petroleum, a mixture of hydrocarbons. If we stop using petroleum, where will we get asphalt for road paving, lubricants, paints, pharmaceuticals, concrete, plastics, ammonia and steel?

18. What would have to happen in order for you to change your mind about climate change?

Made in the USA
Monee, IL
02 November 2026

67387138R00103